Atlassian DevOps Toolchain Cookbook

Recipes for building, automating, and managing applications with Jira, Bitbucket Pipelines, and more

Robert Wen

Alex Ortiz

Edward Gaile

Rodney Nissen

Atlassian DevOps Toolchain Cookbook

Group Product Manager: Preet Ahuja

Publishing Product Manager: Surbhi Suman

Book Project Manager: Ashwin Kharwa

Senior Editor: Mohd Hammad

Technical Editor: Yash Bhanushali

Copy Editor: Safis Editing

Proofreader: Mohd Hammad

Indexer: Hemangini Bari

Production Designer: Jyoti Kadam

DevRel Marketing Coordinator: Rohan Dobhal

First published: July 2024

Production reference: 1200623

Published by Packt Publishing Ltd.

Grosvenor House

11 St Paul's Square

Birmingham

B3 1RB, UK

ISBN 978-1-83546-378-9

www.packtpub.com

To Jill and Charles, who support all my escapades and projects.

– Robert Wen

To Lina, Jeremy, and Isaac, whose unwavering support has made every adventure unforgettable.

– Alex Ortiz

To Meredith and Gracie – you are my co-pilots in this crazy and fantastic life. I wouldn't be the person I am today without your constant support.

– Ed Gaile

To Jennifer, who is always up for another adventure together. You have helped me discover more about myself than I thought there ever could be.

– Rodney Nissen

Contributors

About the authors

Robert Wen is a DevOps consultant at ReleaseTEAM. He specializes in continuous integration, continuous deployment, value stream management, and agile development. He has over 20 years of experience in hardware and software development, working in industries such as defense, aerospace, healthcare, and telecommunications. During his career, he has been a coach as the scrum master for several Agile projects and a trainer for courses on the **Scaled Agile Framework** (**SAFe**). He holds the **SAFe Program Consultant** (**SPC**), Atlassian Certified Expert, Certified CloudBees Jenkins Platform Engineer, GitLab Professional Services Engineer, and Flow Framework Professional titles and certifications.

There are a lot of people in the organizations where I have worked that I would like to thank, without whom this book wouldn't have been possible.

First, many thanks go to my colleagues at ReleaseTEAM, Shawn Doyle, Karl Todd, and the members of the Atlassian practice team.

Second, kudos goes to Atlassian, not only for creating the tools we detail in this book but for fostering a community of enthusiasts to collaborate with, including my co-authors.

Alex Ortiz is an Atlassian educator and administrator at Apetech. He specializes in the administration and utilization of the various Atlassian tools. He has over 10 years of experience in hardware and software development, working for industries such as aerospace and robotics. During his career, he has been an IT administrator for TFS, Jira, Confluence, and Bitbucket. He has also served as a technical program manager for unmanned vehicles and autonomous robots. He is the creator of Apetech Tech Tutorials, a YouTube channel focusing on the Atlassian tools. He is a **Certified Scrum Master** (**CSM**). He has a BS in computer engineering from **California State University, Long Beach** (**CSULB**) and an MS in systems engineering from the **University of Southern California** (**USC**).

I'd like to thank my wife, kids, mother, and father who have believed in me since day one.

A special thanks to Gummy and Carlos for supporting me when everyone else didn't.

And finally, a special thanks to everyone who has ever watched one of my YouTube videos. Without your support, I would have never attempted to even write a book.

Edward Gaile is a principal solution architect at Appfire. He has over 20 years of experience in software development, primarily in the healthcare insurance industry. His specialties include enterprise application design, DevOps, and change enablement optimization. He has been involved in the Atlassian eco-sphere for the past 12 years, including leading the Atlanta Atlassian user community for the past 10 years. He holds multiple Atlassian certifications and a bachelor's degree in industrial engineering from the Georgia Institute of Technology. In his spare time, he is a professional BBQ pit master.

If it takes a village to raise a kid, the same can be said for writing a book.

I would like to thank my co-authors, "King" Bob Wen, Alex Ortiz, and Rodney Nissen. Working together on this project has been extremely rewarding, inspirational, and just plain fun.

I would also like to express my appreciation for the entire Atlassian community, especially fellow community leaders. Truly, they are the most supportive community of amazing people.

Rodney Nissen, better known as *the Jira Guy*, is a blogger who writes regularly on Atlassian tools and cohosts the *The Jira Life* podcast. He is also a DevOps engineer with ReleaseTEAM and has been working on setting up DevOps tools for 10 years. He specializes in integrating various DevOps tools together, Atlassian tools setup and configuration, and using "infrastructure as code" configuration tools to aid in system administration. He is currently an Atlassian Community Leader, Atlassian Certified Expert, and an active participant in the Atlassian Creator Program, holding credentials in Atlassian Jira administration, Atlassian confluence administration, Atlassian Agile development with Jira, and Atlassian system administration.

I'd like to thank my wife, who always supported me and believed I could make this hair-brained scheme work.

I'd also like to thank my many readers, who found my random shouts into the void and gave them a deeper meaning than I could have ever envisioned when I started this journey.

Lastly, my co-authors and co-conspirators, Alex, Bob, and Ed. I've got bad news because, if I have any say in it, we are stuck together now!

Table of Contents

3

Planning and Documentation with Confluence   75

Part 2: Development to Deployment

4

Enabling Connections for Design, Source Control, and Continuous Integration   101

5

Understanding Bitbucket and Bitbucket Pipelines   147

6

Extending and Executing Bitbucket Pipelines   179

7

Leveraging Test Case Management and Security Tools for DevSecOps   217

8

Deploying with Bitbucket Pipelines   245

Part 3: Maintaining Operations

10

Collaborating with Operations through Continuous Deployment and Observability   309

Part 4: Putting It into Practice

13

Putting It All Together with a Real-World Example   431

Preface

DevOps as a movement requires people in teams to examine and change three aspects where they work and interact: people, processes, and tools. Changes for people may include changes in organizational makeup or culture. Process change may include the addition of practices seen in Lean Thinking to ensure continuous delivery. Tool changes allow for quicker turnarounds through automation.

The tools produced by Atlassian can certainly aid our teams with improving processes and fostering automation. Key tools such as Jira and Confluence allow teams to establish standard processes that promote agility and Lean Thinking. Bitbucket Cloud, with its introduction of Bitbucket Pipelines, not only fosters version control but also allows for automation to establish continuous integration and continuous deployment.

But the hallmark of the Atlassian tools isn't the tools themselves; it's the ease to which they can connect to each other and to other tools from different manufacturers. Atlassian's Open DevOps platform provides the ability to create easy interconnections, regardless of vendor.

The focus of this book is not any singular Atlassian product, such as Jira or Confluence. Rather, this book looks to highlight the ease of interconnecting Atlassian tools with other Atlassian tools or other development tools such as GitLab, GitHub, Snyk, and LaunchDarkly. The hope is that by reading this book, you can combine your Atlassian investment with your other investments to create a robust toolchain.

We begin with the assumption that your organization's current state of tools includes tools from Atlassian and other vendors. Because of this, we do not advise reading this book sequentially. We feel the best use of this book is to read the chapters that apply to connecting your tools, without worrying about making an Atlassian-only stack. These incremental changes follow a Lean approach that is key to adopting DevOps.

We conclude the recipes in this book with an example of what an ideal DevOps toolchain looks like in *Chapter 13*. While the sample uses mostly Atlassian tools, it can be made with any combination of tools.

Who this book is for

This book is intended for administrators of Atlassian tools and DevOps and DevSecOps practitioners such as developers and site reliability engineers. Administrators will need to know how to create team structures such as Jira projects, Confluence spaces, and Bitbucket workspaces.

What this book covers

Chapter 1, An Introduction to DevOps and the Atlassian Ecosystem, details a brief history of DevOps and showcases recipes for installing Open DevOps or other Atlassian tools from Jira.

Chapter 2, Discovering Customer Needs with Jira Product Discovery, introduces Jira Product Discovery for product managers to develop and prioritize the seeds of products and features until they are ready for teams to develop.

Chapter 3, Planning and Documentation with Confluence, shows the capabilities of connecting Jira with Confluence to enable up-to-date reporting and documentation.

Chapter 4, Enable Connections for Design, Source Control, and Continuous Integration, contains recipes to connect Jira with other source control and continuous integration tools.

Chapter 5, Understanding Bitbucket and Bitbucket Pipelines, introduces Bitbucket Cloud. We look at the source control features and conclude with an initial look at Bitbucket Pipelines.

Chapter 6, Extending and Executing Bitbucket Pipelines, continues our exploration of Bitbucket Pipelines by seeing how to create and edit its primary control file, `bitbucket-pipelines.yml`.

Chapter 7, Leveraging Test Case Management and Security Tools for DevSecOps, demonstrates recipes to connect to testing and security scan tools and introduces the concept of DevSecOps.

Chapter 8, Deploying with Bitbucket Pipelines, looks at using Bitbucket Pipelines to configure continuous deployment. We demonstrate several examples of deployment to various environments.

Chapter 9, Leveraging Docker and Kubernetes for Advanced Configurations, shows how Docker and Kubernetes can be leveraged by Bitbucket Pipelines to aid in build and deployment.

Chapter 10, Collaborating with Operations through Continuous Deployment and Observability, looks at connecting Jira to tools that provide continuous deployment and monitoring.

Chapter 11, Monitoring Component Activity and Metrics Through CheckOps in Compass, introduces a new Atlassian tool, Compass, to collect and monitor information on deployments and incidents to get a measure of health from project components in a discipline called CheckOps.

Chapter 12, Escalate Using Opsgenie Alerts, demonstrates the use of Opsgenie, which allows for teams to collaborate when an incident occurs.

Chapter 13, Putting It All Together with a Real-World Example, puts everything you have learned from previous chapters together to see how a DevOps toolchain operates.

Chapter 14, Appendix – Key Takeaways and the Future of Atlassian DevOps Tools, finishes with a look at the future and concludes with other tips to get you started on your DevOps journey.

To get the most out of this book

Tool administrators need to know how to create projects in Jira; spaces in Confluence; and workspaces, projects, and repositories in Bitbucket.

Jira users need to know how to create issues.

Confluence users need to know how to create pages.

Bitbucket and other Git server tools (e.g., GitLab and GitHub) users need to know how to perform commits in Git and pull requests/merge requests in the Git server tools of their choice.

Although the Atlassian tools we discuss are cloud-based, **Software-as-a-Service (SaaS)** applications, some actions or features require external resources. We outline these external resources in the following table:

Action/feature covered in the book	External resource requirements
Connect to Moqups (Chapter 4)	Moqups account
Connect to Figma (Chapter 4)	Figma account
Connect to GitLab (Chapters 4 and 10)	GitLab account
Connect to GitHub (Chapter 4)	GitHub account
Connect to Jenkins (Chapter 4)	Jenkins server
Connect to CircleCI (Chapter 4)	CircleCI account
Bitbucket runner – self-hosted (Chapter 6)	Windows, macOS, or Linux machine
Connect to Snyk (Chapters 7 and 13)	Snyk account
Deploy to JFrog using JFrog CLI	Account in the JFrog environment
Deploy to Sonatype Nexus	Account in the Nexus environment
Deploy to AWS S3 (Chapter 8)	AWS account with appropriate role
Deploy to Google Cloud (Chapter 8)	Google Cloud Platform account with appropriate permissions
Deploy to Microsoft Azure (Chapter 8)	Azure account with appropriate permissions
Connect with monitoring tools (Chapter 10)	Datadog account

If you are using the digital version of this book, we advise you to type the code yourself or access the code via the GitHub repository (link available in the next section). Doing so will help you avoid any potential errors related to the copying and pasting of code.

Download the example code files

You can download the example code files for this book from GitHub at `https://github.com/PacktPublishing/Atlassian-DevOps-Toolchain-Cookbook`. If there's an update to the code, it will be updated on the existing GitHub repository.

We also have other code bundles from our rich catalog of books and videos available at `https://github.com/PacktPublishing/`. Check them out!

Conventions used

There are a number of text conventions used throughout this book.

`Code in text`: Indicates code words in text, database table names, folder names, filenames, file extensions, pathnames, dummy URLs, user input, and Twitter handles. Here is an example: "On the page, type `Jira report` in the search bar."

A block of code is set as follows:

```
pipelines:
  default:
    - step:
        script:
          - echo "Running a command"
```

Any command-line input or output is written as follows:

```
kubectl run <my.app> --labels="app=<my.app>" --image=<my.dockerhub.
username>/<my.app>:latest --replicas=2 --port=8080
```

Bold: Indicates a new term, an important word, or words that you see onscreen. For example, words in menus or dialog boxes appear in the text like this. Here is an example: "Select **Insights** to view links to other sources or create such a link yourself by clicking the **Create an insight** button."

> **Tips or important notes**
> Appear like this.

Sections

In this book, you will find several headings that appear frequently (*Getting ready*, *How to do it...*, *How it works...*, *There's more...*, and *See also*).

To give clear instructions on how to complete a recipe, use these sections as follows.

Getting ready

This section tells you what to expect in the recipe and describes how to set up any software or any preliminary settings required for the recipe.

How to do it...

This section contains the steps required to follow the recipe.

How it works...

This section usually consists of a detailed explanation of what happened in the previous section.

There's more...

This section consists of additional information about the recipe in order to make you more knowledgeable about the recipe.

See also

This section provides helpful links to other useful information for the recipe.

Get in touch

Feedback from our readers is always welcome.

General feedback: If you have questions about any aspect of this book, mention the book title in the subject of your message and email us at `customercare@packtpub.com`.

Errata: Although we have taken every care to ensure the accuracy of our content, mistakes do happen. If you have found a mistake in this book, we would be grateful if you would report this to us. Please visit `www.packtpub.com/support/errata`, selecting your book, clicking on the Errata Submission Form link, and entering the details.

Piracy: If you come across any illegal copies of our works in any form on the Internet, we would be grateful if you would provide us with the location address or website name. Please contact us at `copyright@packtpub.com` with a link to the material.

If you are interested in becoming an author: If there is a topic that you have expertise in and you are interested in either writing or contributing to a book, please visit `authors.packtpub.com`.

Share Your Thoughts

Once you've read *Atlassian DevOps Toolchain Cookbook*, we'd love to hear your thoughts! Scan the QR code below to go straight to the Amazon review page for this book and share your feedback.

https://packt.link/r/1835463789

Your review is important to us and the tech community and will help us make sure we're delivering excellent quality content.

Download a free PDF copy of this book

Thanks for purchasing this book!

Do you like to read on the go but are unable to carry your print books everywhere?

Is your eBook purchase not compatible with the device of your choice?

Don't worry, now with every Packt book you get a DRM-free PDF version of that book at no cost.

Read anywhere, any place, on any device. Search, copy, and paste code from your favorite technical books directly into your application.

The perks don't stop there, you can get exclusive access to discounts, newsletters, and great free content in your inbox daily

Follow these simple steps to get the benefits:

1. Scan the QR code or visit the link below

https://packt.link/free-ebook/978-1-83546-378-9

2. Submit your proof of purchase
3. That's it! We'll send your free PDF and other benefits to your email directly

Part 1: Beginning the Cycle

This book is an exploration of DevOps as a whole and where tools fit. Of particular interest are the connections that Atlassian tools can make with each other and with tools from other manufacturers.

We set the stage for our exploration of DevOps and toolchains by understanding the background and context of why this approach is needed. Once we have discovered the "why" of DevOps, we shift attention to Jira and the connections it can make through the Open DevOps platform. We discover how to configure Jira to connect with other tools using this platform.

Before any implementation begins, developers have to work with product managers, stakeholders, and perhaps even the customer to determine the need, viability, and feasibility of a new product. Jira Product Discovery aids this effort of determining which ideas should be acted on by providing a location for collecting, tracking, elaborating, and comparing different ideas for products and features. Those ideas that are ready can be linked to a Jira issue to monitor development.

As we start development, we need to plan our efforts and ensure our efforts are on track. We also want to ensure that the development efforts link to the overall big picture of the product and its promise of value. To that end, we will connect Jira to Confluence so that Confluence can provide the necessary project documentation.

This part has the following chapters:

- *Chapter 1, An Introduction to DevOps and the Atlassian Ecosystem*
- *Chapter 2, Discovering Customer Needs with Jira Product Discovery*
- *Chapter 3, Planning and Documentation with Confluence*

1

An Introduction to DevOps and the Atlassian Ecosystem

DevOps has been a driving force for improvement in product development since its inception in 2009. As technology moved to internet-based, off-premises cloud environments, DevOps allowed people working in development and operations a way to collaborate, enabling quicker and more stable design, packaging, deployment, and maintenance of products. A key component of this is the adoption of automation that ensures these processes run smoother.

In this chapter, we will look at the DevOps movement and the role that automation plays in its success. We will then look at the **Open DevOps** platform from **Atlassian** that allows easy connection between Jira, other Atlassian tools such as Confluence and Bitbucket, as well as third-party tools such as LaunchDarkly, which allows you to release products through feature flags, and Snyk, which performs security scanning.

To start our journey into Open DevOps, we will examine the steps needed to enroll in Atlassian's cloud environment to obtain trial versions of Jira and connections through Open DevOps. To accomplish this, we will cover the following recipes:

- Creating the Open DevOps toolchain from scratch
- Creating a new Atlassian Cloud site with Jira only
- Creating a Jira project
- Connecting Confluence
- Connecting Bitbucket

An introduction to DevOps

By 2009, developers began to adopt Agile product development methods. They started with small deliveries in incremental cycles, gathering customer feedback and using it to inform future development cycles. They gradually created a product that their customers would want.

A bottleneck would soon form from delivering changes to the operations part of the organization. While rapid development of value is the priority of development teams, operations teams are charged with maintaining the stability of the environment. Anything that would diminish that stability would mean a potential loss of revenue. Typically, that meant seeing any new change as risky and allowing any changes to be released in specific maintenance windows. These windows only ended up increasing the risk of further downtime.

Changes would soon emerge. During the *O'Reilly Velocity 2009 conference*, John Allspaw and Paul Hammond of Flickr gave a talk titled *10+ Deploys a Day – Dev and Ops Cooperation at Flickr* that described the different practices that allowed them to perform multiple deployments in a single day, in contrast to many organizations that were struggling to perform a deployment in a single year. Patrick Debois, after watching the aforementioned talk and other talks on the importance of collaboration between development and operations, and being intrigued by the concept of Agile system administration following a conversation with Andrew Schafer the year before, decided to organize a conference in Ghent, Belgium, to address the topic. To emphasize this need, Debois added the common abbreviations of *development* and *operations* and combined them in the name of his conference – **DevOpsDays**.

The momentum of the first DevOpsDays conference continued to social media. Discussion continued on Twitter (now known as X), where the topic was identified through the hashtag #DevOps. Additional DevOpsDays conferences were organized in various cities throughout the world. The movement caught the attention of Gartner, who, in 2011, predicted that DevOps would soon be adopted by enterprises. This ensured that DevOps moved from an underground movement to a mainstream idea to implement in business.

Now that the idea of collaboration between development and operations has shown benefits and proven to be popular, one key question is how successful DevOps implementations are started. A key model to that approach is identified in the next section.

The CALMS/CALMR approach

In preparation for the *2010 DevOpsDays conference*, John Willis and Damon Edwards were asked the same question about the burgeoning DevOps movement – how do we implement DevOps? In other words, what were the important factors for a successful DevOps implementation?

They came up with an acronym, **CAMS**, where each letter identified a key component for a successful DevOps approach. **C** stood for **Culture**, **A** represented **Automation**. **M** was for **Measurement**, and **S** stood for **Sharing**.

When writing in *The DevOps Handbook*, Jez Humble elaborated on the CAMS model. He added an L, which stood for **Lean**. With this addition, CAMS became **CALMS**.

Scaled Agile Inc., when adopting DevOps into the **Scaled Agile Framework (SAFe)**, modified the approach. Reasoning that the ideal DevOps culture was one of shared responsibility, Scaled Agile removed the S for Sharing and replaced it with an **R**, which stood for **Recovery**. CALMS became **CALMR** when practicing DevOps in SAFe.

Let's take a look at each letter of the CALMS and CALMR approaches and see how they fit in with the adoption of Atlassian tools and the Open DevOps platform.

Culture

The management expert Peter Drucker is credited with saying, *"Culture eats strategy for breakfast,"* underlying the impact that culture, as a common thread, can have on uniting individuals in a group, from as small as a team, to as large as nations. So, if the correct culture can drive organizations to desired results, what is the correct culture? To find the answer, we turn to a sociologist named Ron Westrum.

In 1988, Ron Westrum organized a study to measure the safety of medical teams. He organized the teams into three different types of cultures:

- Pathological
- Bureaucratic
- Generative

A pathological culture is one that's leader-driven. Motivation mainly comes through fear and threats from the leaders to accomplish the (leader's) goals.

A bureaucratic culture has safety mechanisms through rules and standards, but these can then be used to protect the members of the group from outsiders.

A generative culture, in contrast, is focused on aligning with a common mission. Information is freely shared with whomever, whether they are a member of the group or not, irrespective of whether they can play a role in the success of the mission.

Westrum found in his initial study that medical teams that possessed a generative culture also possessed alignment to their mission, an awareness of things that impeded the mission, and gave empowerment to any individual to make changes to remove those impediments. This allowed organizations to easily make long-lasting improvements to the system to improve the teams even more.

These benefits are not limited to medical teams. In the book *Accelerate – Building and Scaling High Performing Technology Organizations*, Nicole Forsgren, Jez Humble, and Gene Kim investigated recommending DevOps practices to teams to see how effective they were. They surveyed development teams and found that those teams that had a generative culture produced higher levels of performance for software delivery and experienced higher levels of job satisfaction.

A change in culture, while possible, is typically the last change that occurs after structural and behavioral changes. Atlassian does provide tools that assist in the change of structure and behavior to move teams to a generative culture. These tools that assist cultural change are available for free as the Team Playbook (https://www.atlassian.com/team-playbook). However, going into detail about these tools is beyond the scope of this book.

Automation

When people think of DevOps, the first thing that comes to mind is automation. In the talk given by Allspaw and Hammond at the O'Reilly Velocity 2009 conference, one of the key factors of success mentioned included the use of automated infrastructure and a common version control system for both development and operations.

These days, key automation is done by successfully linking tools together to form a toolchain. Each activity in development and operations has at least one tool associated with it.

A diagram that shows development activity with associated Atlassian tools is shown in the following figure:

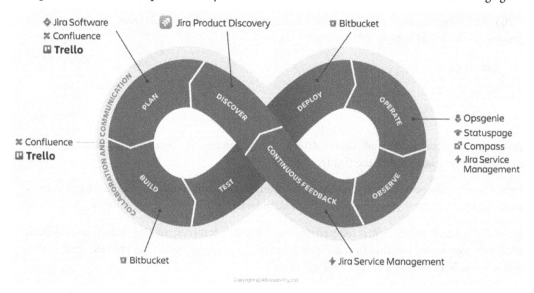

Figure 1.1 – DevOps phases with Atlassian tools

This book will demonstrate in its chapters how to establish a toolchain with the Atlassian tools illustrated in the preceding figure.

Lean

The system created by automation can only work well if its demand does not exceed its capacity. To ensure this balance, we will look at applying lean thinking practices, initially developed as part of the Toyota Production System. Practices include the following:

- Make all work visible
- Limit the **Work-in-Progress (WIP)**
- Keep batch sizes small
- Monitor queues

We will see the relationships between the size of the work (the queue length), the time it takes to process the work (the cycle time), and how long until we see the results (the wait time) from queueing theory. A key formula in this is Little's law, expressed here:

$$L = \lambda W$$

L signifies the queue length. The Greek letter *lambda* represents the throughput the team has in processing the work. W represents the wait time for finished work.

An additional equation called Kingman's formula tells us that there's a direct correlation between the cycle time, utilization, and variability of the items of work the team must complete. This formula (often called the *VUT equation*) is shown here:

$$E(w) \approx V \times U \times T$$

In this equation, *E(W)* represents the wait time. It is shown as equivalent to the product of *variability (V)*, *utilization (U)*, and *cycle time (T)*.

Jira has features that allow us to display the work to do through Kanban boards, also allowing us to limit the WIP through column constraints. Wait times and cycle times can also be calculated from the metrics collected.

Measurement

In judging the effectiveness of our efforts to develop, release, and maintain our products, we need to ask three key questions:

- Are we on track to deliver the solution?
- What is the health of all of our operating environments (test, stage, and production)?
- Do our customers think we have developed the right solution?

We saw when looking at *lean* practices in the previous section that we needed to pay attention to the following metrics to evaluate whether we operate in a state of flow:

- The cycle time
- The WIP
- The throughput
- Any impediments or blockers to progress

In looking at the condition of the operating environments, we turn to the discipline of observability, where we collect all aspects of the environments from infrastructure to applications. The typical measurements for observability include the following:

- Logs
- Call/execution traces
- Metrics

Measurements that can indicate whether our customers receive the proposed value from new product features can be tricky to determine. Vanity metrics may exist that occur naturally and indicate good trends, but after thorough analysis, they do not produce actionable answers. Metrics that have proven reliable in measuring customer sentiment include the following:

- *Pirate* metrics (Acquisition, Activation, Retention, Referral, and Revenue) devised by Dave McClure, a look at ideal customer behavior toward a product or feature, with measurements of that behavior
- Google HEART Metrics, used by Google UI/UX to gauge user preferences in terms of Happiness, Engagement, Adoption, Retention, and Task Success
- **Fit for Purpose** (**F4P**) devised by David J. Anderson and Alexei Zheglov, which measures whether a customer's needs are met, as outlined in the book *Fit for Purpose – How Modern Businesses Find, Satisfy, and Keep Customers.*

Atlassian applications can collect the needed metrics or easily integrate with dedicated metric collection tools. Jira is a proven platform to collect the metrics needed for lean. Out-of-the-box reports, as well as the easy application of third-party marketplace apps, allow us to collect and analyze required metrics. Open DevOps allows easy integration with observability tools such as those offered by Datadog and New Relic.

Sharing

Once we have collected the measurements, we need a way of easily displaying the information to all who need it to foster a generative culture.

The Atlassian tools described in this book provide a good basis to foster information sharing and transparency. Jira can produce charts and reports that can be shared on a dashboard. Other displays related to application health can be collected and displayed using Compass.

Recovery

In any DevOps implementation, we want to devote time and attention to planning contingency steps to take if a release causes an outage in the production environment. DevOps changes the operating model where both development and operations, either acting together or with specialized site reliability engineers, must collaborate to answer the following questions:

- How do we reduce the risk when we release?

- What mitigation steps should be designed to limit the outage time, should it occur?

- If an outage happens, what procedures should we follow?

Atlassian tools may form the answer to these problems. Feature flag tools such as those provided by LaunchDarkly, easily integrate with Jira through Open DevOps. Compass provides *early warning* capabilities if an outage appears imminent. Opsgenie allows development and operations to collaborate in an outage to bring about a swift resolution.

We have seen the key pillars of a DevOps approach and how the Atlassian tools help foster them. In the next section, let's examine the possible technologies that impact DevOps implementation and where Atlassian tools can help.

Technology considerations for DevOps

The adoption of the DevOps approach was encouraged by emerging technologies that made it easier and faster to deploy, release, and maintain products. The changes in technology include the following:

- **Continuous Integration/Continuous Deployment (CI/CD)** pipelines

- **Infrastructure-as-Code (IaC)**

- Cloud environments

- Containers/Kubernetes

Let's take a closer look at these factors.

CI/CD pipelines

At the time of Allspaw and Hammond's talk at the O'Reilly Velocity 2009 conference, CI tools were often used to make a *daily* build or a build of software that gathered the commits of that day. After creating the build, the CI tools would run automated tests, and report the success or failure of that operation.

Allspaw and Hammond took their CI tool further. If a build passed all its tests, the CI tool would allow you to move the build artifacts to either a test or staging environment for further testing, or to the production environment to prepare for release or *go-live*.

The extension of CI came to be known as CD. Automating deployment by using a CI/CD tool allows for consistent deployments because steps aren't forgotten or skipped. Testing of deployment further ensures proper function and that the behavior seen is the desired behavior.

As tooling moved from just automated building to incorporating all steps of CI and CD from a single trigger, version control tools began to offer their own pipeline capabilities, controlled by a text file in the **YAML (YAML Ain't Markup Language)** format. This facilitated a mini-movement called GitOps where build, testing, integration, packaging, and deployment all started from a single Git commit as a trigger.

IaC

As CI expanded to include CD, deployment into environments was about to be made easier. New tools such as Ansible, Chef, Terraform, and Puppet allowed for the definition of an ideal infrastructure as specified by text files.

By running the infrastructure tools with the text files as input, there would be a consistent way of creating environments, whether they were used for testing, staging, or production. This consistency helped in keeping environments similar, ensuring the same testing results no matter which environment was used and preventing configuration drift, where problems were seen when environments weren't similar.

Cloud environments

While CI/CD pipelines and IaC tools could be applied to many physical platforms, from cyber-physical systems that became known as the **Internet of Things (IoT)** to physical servers, DevOps success has become associated most closely with the rise of cloud environments.

Cloud environments are created with virtual machines available from a vendor and are accessible through the Internet. Creation and disposal of the virtual machines can happen within minutes, allowing for dynamic and flexible setups that could be provisioned on demand.

Containers and Kubernetes

The popularity of cloud environments and IaC has encouraged further thought as to how to package a software application and propagate that to multiple test, staging, and production environments.

Containers have a long history as an approach to isolate processes and their resources, first in Unix and then in Linux. In 2013, Docker was the first company to introduce not only its standard of containers but also a way of managing them. The standard provided by Docker is the most prevalent one used today in describing, creating, and managing containers.

What is a container? If we start with the application we create and how its resources are allocated on a physical computer server, we can see that its code, libraries, and data occupy some of the server's memory, as illustrated in the following figure:

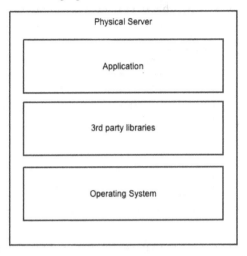

Figure 1.2 – A packaged application in a physical server

Virtual machines allow a physical server to host multiple instances of **Virtual Machines** (**VMs**). Each VM can have its own individual set of operating systems, applications, and accompanying libraries. The VMs are managed by a hypervisor, which is an application process run on the physical server. This arrangement is shown as follows:

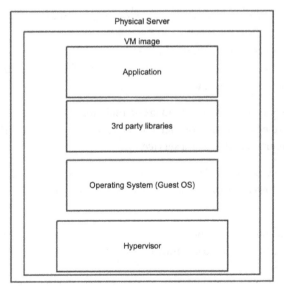

Figure 1.3 – VM and application packaging in a physical server

Docker or other container management systems allow for more efficient resource management. The container requires only resources for the application and any dependent third-party libraries. Any OS-level resources that are required are managed through a managing application. In Docker's case, the managing software is Docker Engine. This arrangement is outlined as follows:

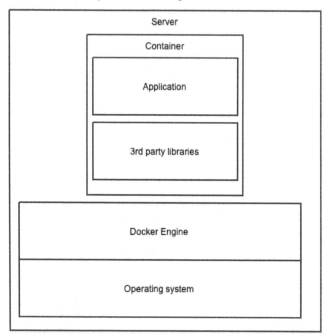

Figure 1.4 – Container and server packaging

Note that the server that hosts Docker Engine and its containers can be a physical server or a VM. This allows for portability, where the only artifact that can move from environment to environment is the definition of the container, known as the container image.

With the portability of containers, developers can create their application and deploy it into an artifact that can be version-controlled and easily transferred to testing or production. The application under development can be compartmentalized as a service.

Kubernetes emerged from Google in 2015 as a way of creating and managing clusters composed of containers. The clusters allowed for the dynamic creation of Pods from containers to make flexible and resilient services or microservices.

Now that we have seen the underlying tenets of DevOps and the main facets of technology that drive its acceptance and popularity, it's time to examine how the tools from Atlassian can form a toolchain for a DevOps approach.

Creating an Open DevOps toolchain from scratch

The first link to a toolchain using Atlassian tools and Open DevOps is to set up an organization in the Atlassian Cloud. The first Atlassian product in the organization will be Jira.

In this recipe, we will explore how to create the organization and an Atlassian Cloud site with Jira installed.

How to do it...

Let's go to the Atlassian website to create an Open DevOps environment with multiple connected tools:

1. Atlassian's Open DevOps solution is described at `https://www.atlassian.com/solutions/devops`. This site allows you to see what Open DevOps is, what third-party integrations make up part of the solution, and what DevOps best practices Open DevOps plays a role in.

2. From this page, you can request the creation of a new Atlassian Cloud site with the Jira product, installed by selecting the **Try for free** button on the page, as shown in the following figure:

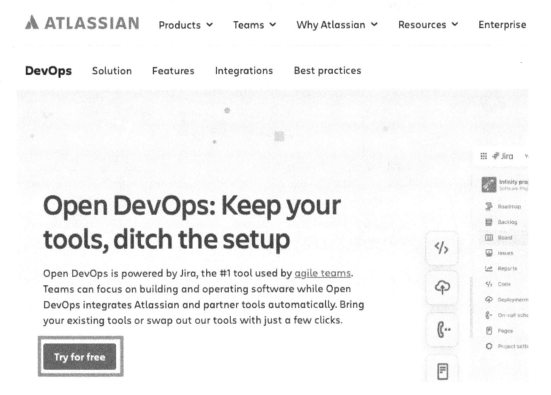

Figure 1.5 – Atlassian's Open DevOps page with the Try it free button

3. The page will open, allowing you to connect an existing Bitbucket workspace or create a new one by typing its name. This panel is shown in the following screenshot:

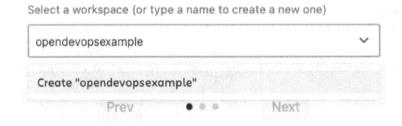

Let's get started

Integrate Jira, Confluence and Opsgenie with Bitbucket for free, then upgrade or add more third-party tools later.

Select a workspace (or type a name to create a new one)

opendevopsexample ⌄

Create "opendevopsexample"

Prev ● ◦ ◦ Next

NO CREDIT CARD REQUIRED

Figure 1.6 – Selecting a Bitbucket workspace

4. Create a new Bitbucket workspace to tie into your toolchain. This is the default option when creating a new toolchain. You can also select an existing Bitbucket workspace if you have one. After you've made your choice, select **Next**.

5. The next panel prompts you to name the new Atlassian Cloud suite. The text before **.atlassian. net** must be unique. This panel is depicted as follows:

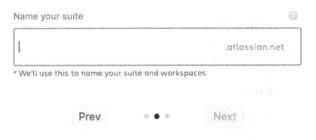

Figure 1.7 – Naming your Atlassian Cloud suite panel

6. Name your Atlassian Cloud site and press **Next**.

7. A new panel should indicate success, prompting you to go to Jira on your cloud site and create a project, as seen in the following figure:

Figure 1.8 – The Open DevOps Site success panel

8. During the entire process, the panel has specified that no credit card is needed, and we haven't had to include payment information. What gets created on our Atlassian Cloud site are the free plans for Jira, Confluence, and Bitbucket and the DevOps plan for Opsgenie, which is free for five users.

Following this recipe has allowed for the creation of an entire toolchain, starting with Jira, Bitbucket, Confluence, and Opsgenie.

Creating an Atlassian Cloud site with Jira only

A more modular approach than creating an Atlassian Cloud site with multiple products can be to create a new Atlassian Cloud site and install Jira only. Other installations of other Atlassian products can be done on an as-needed basis.

To do this, we will go to a different landing page and follow the prompts, as seen in the following recipe steps.

How to do it...

To install Jira only on a new Atlassian Cloud site, follow these alternate instructions:

1. Atlassian's product page for Jira is located at `https://www.atlassian.com/software/jira`. Select the **Get it free** button, as shown in the following figure:

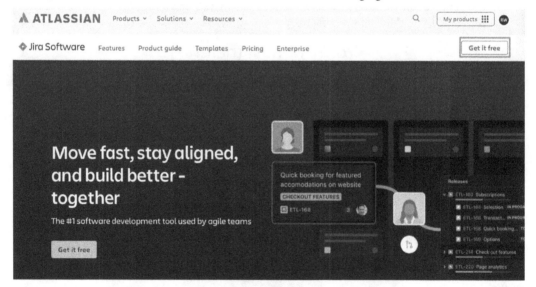

Figure 1.9 – Selecting the free plan from the Jira Products page

2. On the next panel, if you're logged into an Atlassian account, it will reveal that account and prompt you to enter the name of the new Atlassian Cloud site.

Welcome back, Robert

Work email

Sign in with a different Atlassian account

Your site 🌐

.atlassian.net

By clicking below, you agree to the Atlassian Cloud Terms of Service and Privacy Policy.

Agree

⚠ ATLASSIAN

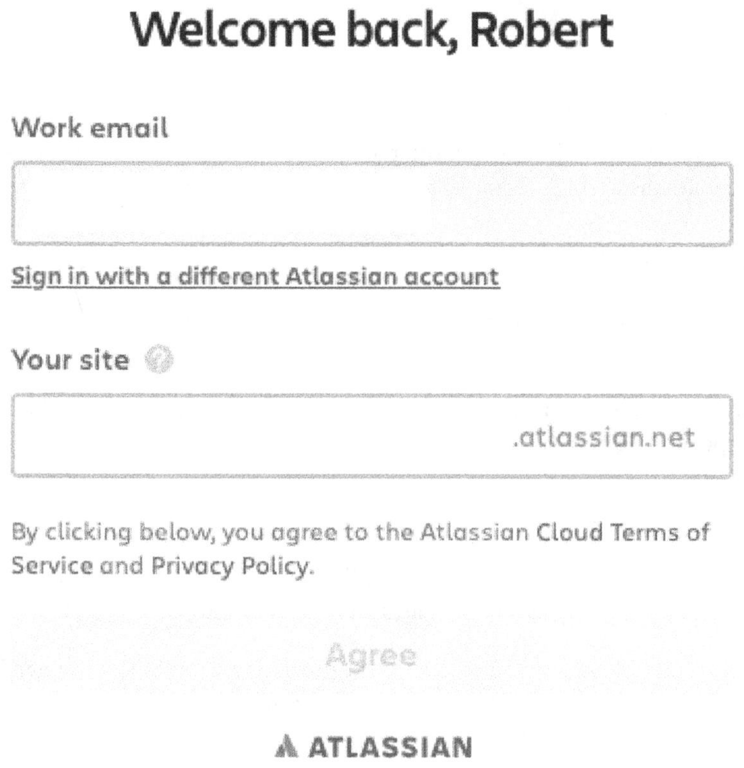

Figure 1.10 – Creating a new Atlassian Cloud site

3. As seen in the preceding figure, fill in the Atlassian account and the desired name for your Cloud site and click **Agree**.

Following the instructions will create an Atlassian Cloud site with the name you have specified, granting you administrator privileges. The next section looks at the scenario where you have to create an Atlassian account if you don't have one.

There's more...

If you don't have an Atlassian account or haven't logged into your Atlassian account, you will be taken to a different page to create a free Atlassian account:

1. You can enter a work email or use common accounts with OpenID service providers, such as Google, Microsoft, Apple, or Slack.

2. Once the account is created, Atlassian will prompt you to create an Atlassian Cloud site with a free plan of Jira installed, as seen before.

Creating a Jira project

So far, you have created an Atlassian Cloud site with only Jira or Jira with Confluence, and connections to Opsgenie and Bitbucket.

We will start the process of creating our toolchain by setting up the Jira projects that will use the toolchain. Let's look at doing this now.

How to do it...

In the following steps, you will create a Jira project that not only captures the work done by the development team but allows integrations with artifacts from other tools:

1. Go to the Atlassian Cloud site you created in the previous recipes in this chapter. You may be prompted to create the site's first Jira project by selecting a project template, as shown in the following figure:

Select a template for your first project

If you're not sure what to choose, don't worry. You can quickly create a new project if this one's not right for you.

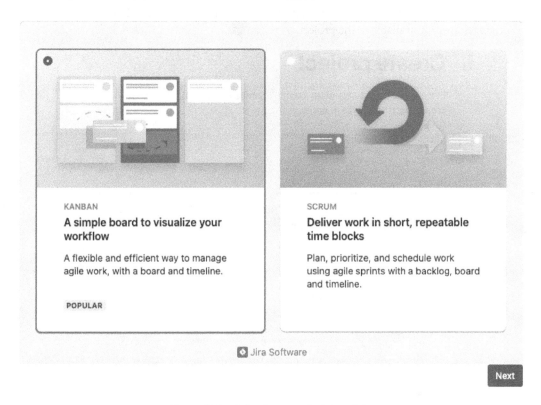

Figure 1.11 – Selecting a project template

2. Select a project template that matches the Agile methodology your team wants to practice (**KANBAN** or **SCRUM**) and press **Next**.

> **Tip**
> In general, teams that look to develop on a cadence with frequent stakeholder feedback opt for Scrum, and teams that look to monitor and track a steady flow of work will choose a Kanban project.

3. The following page allows you to select other options, such as a name and whether the project is company-managed or team-managed. Company-managed Jira projects will have project elements such as custom fields, workflows, notifications, and permissions set up by Jira administrators. Team-managed projects allow all customizations to be done by team administrators, but these changes are only available to the project:

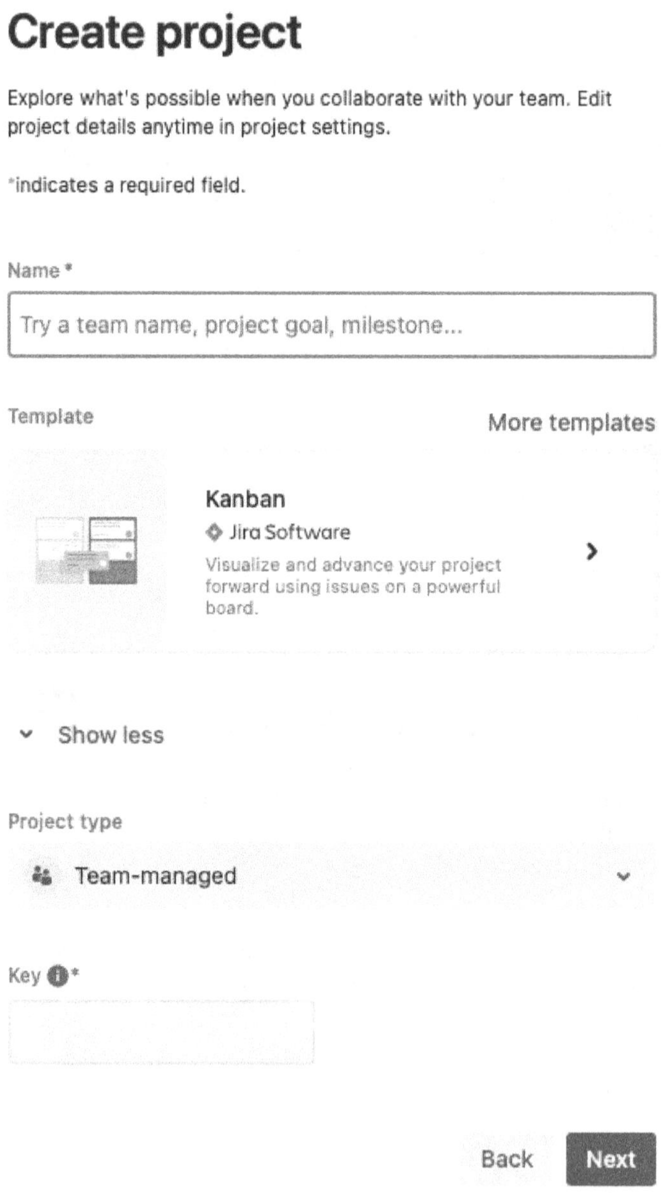

Create project

Explore what's possible when you collaborate with your team. Edit project details anytime in project settings.

*indicates a required field.

Name *

```
Try a team name, project goal, milestone...
```

Template   More templates

Kanban
◆ Jira Software
Visualize and advance your project
forward using issues on a powerful
board.   >

∨   Show less

Project type

👪 Team-managed   ∨

Key ⓘ *

```

```

Back   **Next**

Figure 1.12 – Creating Jira project details

4. If you created an Open DevOps platform on your Atlassian Cloud site, it would automatically create the other artifacts such as Confluence spaces, Bitbucket repositories, and Opsgenie teams. The next panel, as depicted in the following figure, shows this:

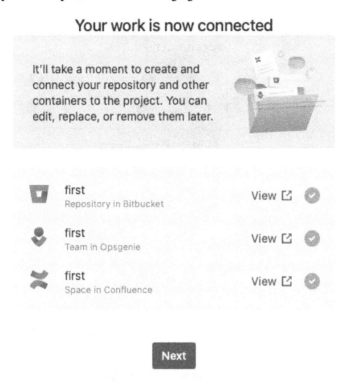

Figure 1.13 – The open DevOps artifacts created

5.   If you created an Atlassian Cloud site with only Jira, you won't see the preceding screen (don't worry; we'll show you how to connect the other tools in other recipes in this chapter). Instead, you will see other tools you can connect to your Jira project, as shown in the following figure:

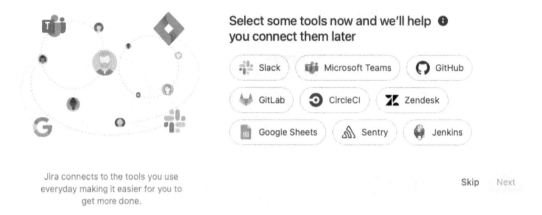

Figure 1.14 – Connecting other tools

6.   You can select the tools to connect and press **Next** to integrate with other tools or select **Skip** to do the integrations later.

By following the preceding steps, you should have a Jira project ready to connect to the necessary tools by means of the Issues or work artifacts, stored in Jira.

The following recipes allow you to make connections when the Open DevOps platform wasn't selected to create the Atlassian Cloud site, or when the Jira product and projects were already created.

Connecting Confluence

Confluence is Atlassian's second-oldest tool. Previously, it has complemented Jira as the repository of documentation artifacts for development activities recorded in Jira.

Confluence works by organizing individual pages into **spaces**. Spaces can be set up for any number of purposes, such as a knowledge base or product documentation.

The first step will be to add Confluence to the Atlassian Cloud site. Once added, a new space can be created for the new Jira project.

Getting ready

Different administrators set up different things for Confluence. Let's take a brief look at the various levels of administration for Atlassian Cloud.

Organization administrators are responsible for the overall organization and the Atlassian Cloud sites contained within. They set up users to access specific Atlassian products and add products to Atlassian Cloud sites.

Confluence administrators are responsible for creating the spaces on an instance of Confluence and managing the permissions of users for those spaces, based on roles.

We will identify what level of administration is needed for each part of this recipe.

How to do it...

We will approach this recipe in two parts – the first part is where we add Confluence to the Atlassian Cloud site that also has Jira. The second part of this recipe is creating a Confluence space and connecting it to a Jira project.

Adding Confluence to the Atlassian Cloud site

Let's look at adding Confluence to the same Atlassian Cloud site where we added Jira only in the *Creating a new Atlassian Cloud site with Jira only* recipe:

1. As the Organization admin, go to https://start.atlassian.com to find the Switcher icon in the top-left corner.

Figure 1.15 – Switching Atlassian applications on the start page

2. Clicking on the icon, as seen in the preceding figure, reveals the application options. Select **More Atlassian products**, as shown in the following figure:

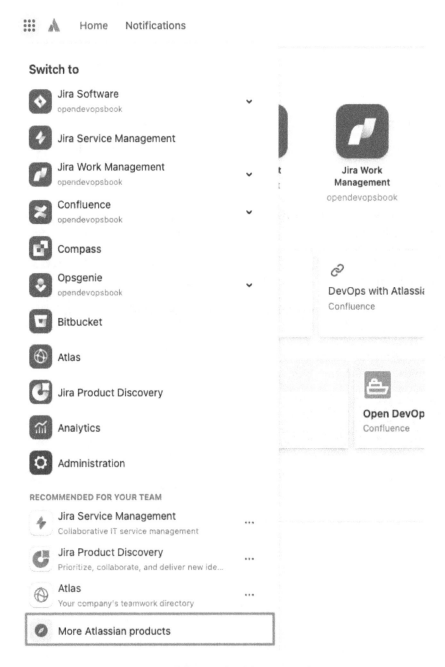

Figure 1.16 – Selecting the Atlassian product

3. On the page that opens, look for the option for Confluence and press **Try it now**.

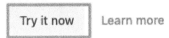

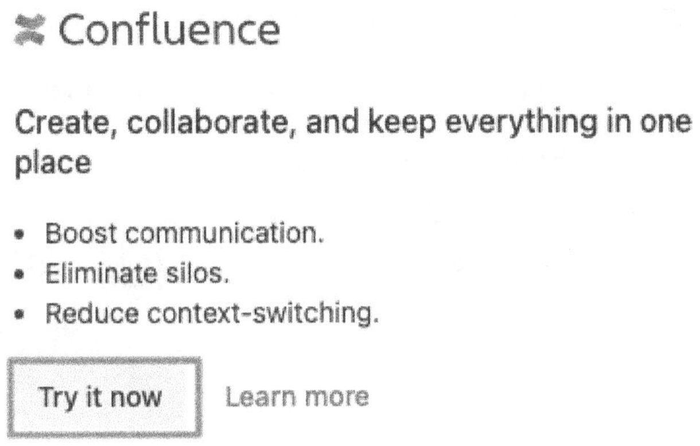

Confluence
Create, collaborate, and keep everything in one place
Boost communication.
Eliminate silos.
Reduce context-switching.
Try it now
Learn more

Figure 1.17 – Trying Confluence

4. Type in the site name where you want to add Confluence and select **Continue**, as shown in the following figure:

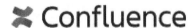

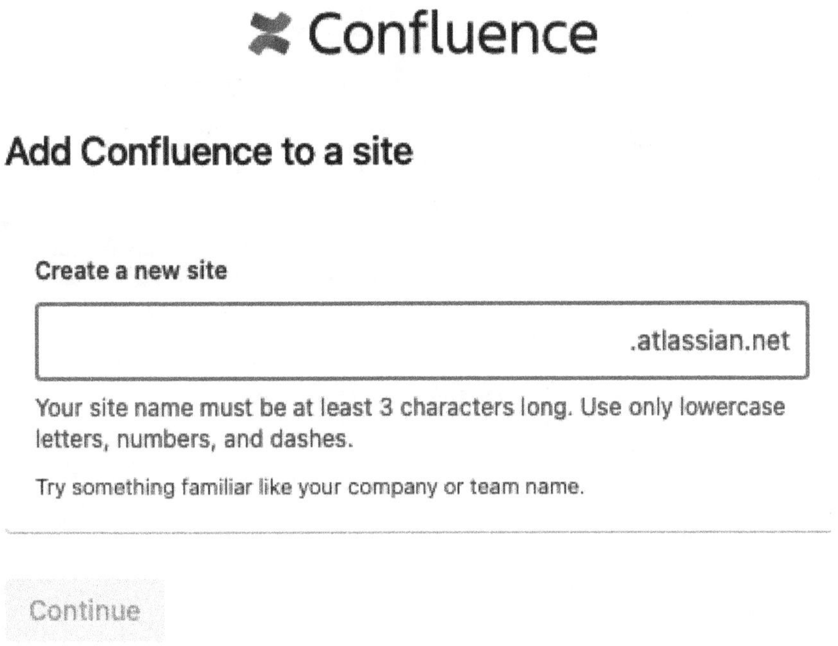
Confluence
Add Confluence to a site
Create a new site
.atlassian.net
Your site name must be at least 3 characters long. Use only lowercase letters, numbers, and dashes.
Try something familiar like your company or team name.
Continue

Figure 1.18 – Add Confluence to a site

The results of these instructions should be the Confluence product added to your Atlassian Cloud site.

Now, let's take a look at filling Confluence with data, in the form of spaces and pages.

Creating a Confluence space and connecting it to the Jira project

Let's start adding data to Confluence by creating our first space:

1. As an organization admin or a Confluence admin, from the **Home** selection on the menu bar, click on the + symbol in the **SPACES** section, as shown in the following figure:

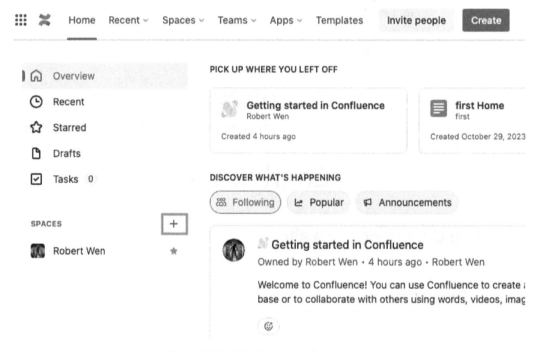

Figure 1.19 – Selecting + to create a space

2. On the following page, select a space template or leave it as a **Blank** template. Click **Next**.

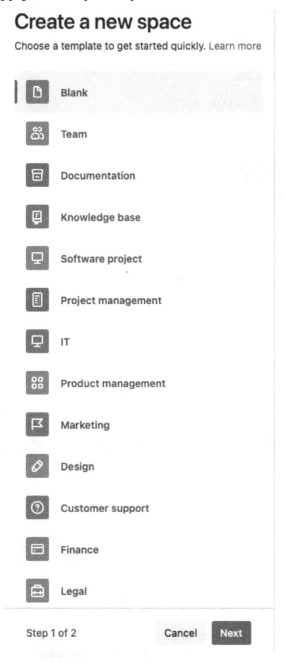

Figure 1.20 – Selecting a Confluence template

3. Give the template a name. You can also specify the space's key. Click on **Create space** to complete the process.

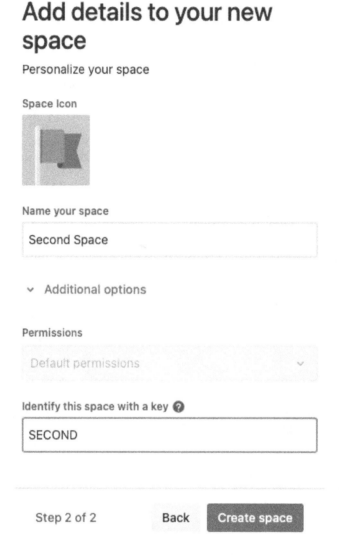

Add details to your new space

Personalize your space

Space Icon

Name your space

Second Space

⌄ Additional options

Permissions

Default permissions   ⌄

Identify this space with a key ❓

SECOND

Step 2 of 2   Back   Create space

Figure 1.21 – Completing the space creation

4. As a Jira project admin, Jira admin, or organization admin, navigate to the desired Jira project. Select **Project pages**. Then, select **Connect to Confluence** or use the arrow icons to view the Confluence spaces or pages you want to connect.

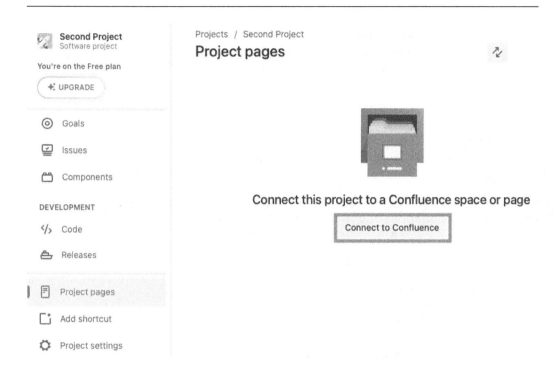

Figure 1.22 – Connecting the Confluence space

5.   Select the space created and click **Connect**, as shown in the following figure:

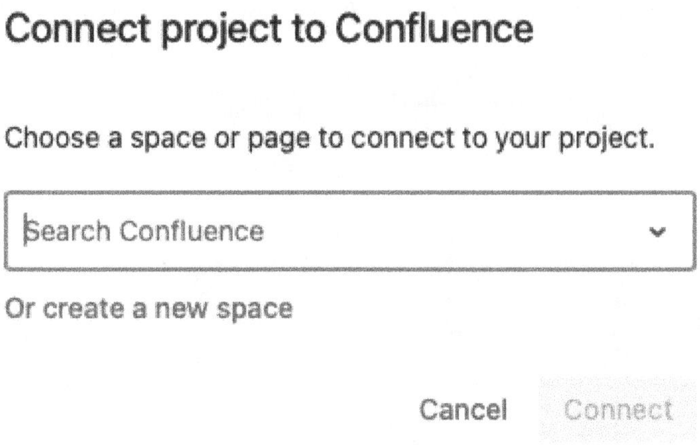

Connect project to Confluence

Choose a space or page to connect to your project.

Search Confluence

Or create a new space

Cancel   Connect

Figure 1.23 – Completing the connection to a space

You have successfully connected Confluence to Jira by defining a space and linking it to the project. Further applications of the benefits that can arise from this integration will be seen in the recipes in *Chapter 3*.

Connecting Bitbucket

Currently, Bitbucket maintains a separate location from the other tools at the Atlassian Cloud site, at `https://bitbucket.org`. On Bitbucket, there are by default administrators and users in a developers group.

On the Bitbucket side, a member of the Administrators group must create the repository or repo. This repo may be grouped within other repositories in a Bitbucket project. The top level in Bitbucket is a workspace. This may be analogous to the top level of an organization.

On the Jira side, we need a Jira Project admin to connect the repository to the Jira project.

Let's take a look at the steps involved.

How to do it...

Let's look at the steps involved in connecting Bitbucket.

Creating a repository in Bitbucket

Implement the following steps to create a repository in Bitbucket:

1. As a Bitbucket administrator, select the **Repositories** menu bar and click the **Create repository** button.

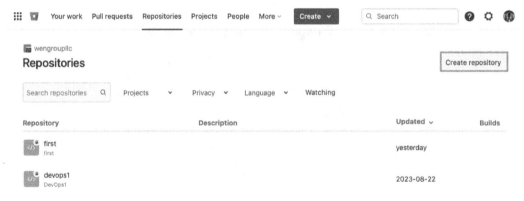

Figure 1.24 – Creating a Bitbucket repository

2. In the page that follows, fill in the repository name and select the project for the repo (these are the only mandatory items) and other desired details. Click the **Create repository** button when finished.

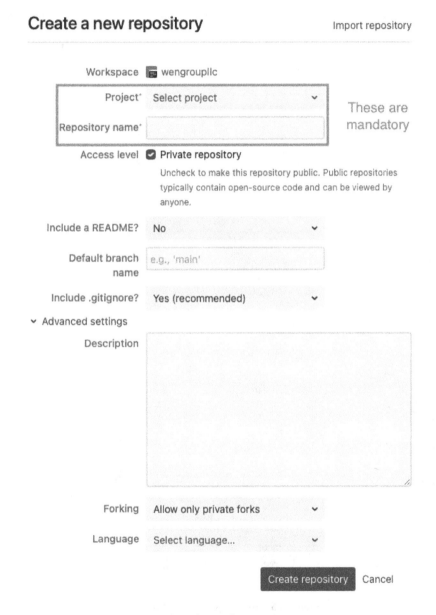

Figure 1.25 – Completing the repository creation

Now that we have a Bitbucket repository, let's connect it to a Jira project.

Connecting a repository to a Jira project

Follow these steps to integrate Bitbucket data from the repository into a Jira project:

1. As a Jira admin, from the Jira project, select **Project settings** from the options on the left.

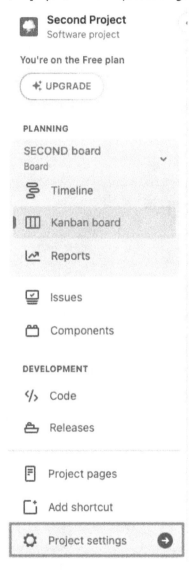

Figure 1.26 – Selecting Project settings

2. Select **Development tools**, as shown in the following figure:

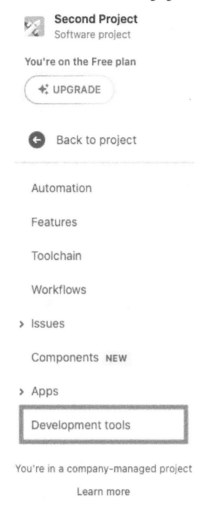

Figure 1.27 – Selecting Development tools

3. By default, Bitbucket is shown to be connected to Jira on the **Development tools** page. We can make adjustments to the connection by selecting the **Connect** pulldown and choosing **Bitbucket Cloud** from the options.

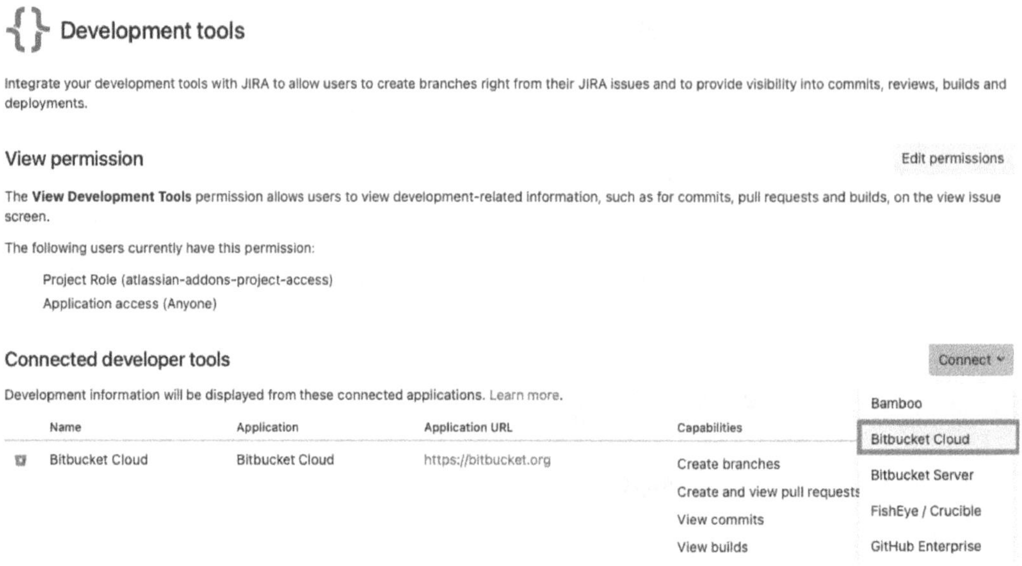

 Development tools

Integrate your development tools with JIRA to allow users to create branches right from their JIRA issues and to provide visibility into commits, reviews, builds and deployments.

View permission

Edit permissions

The **View Development Tools** permission allows users to view development-related information, such as for commits, pull requests and builds, on the view issue screen.

The following users currently have this permission:

Project Role (atlassian-addons-project-access)

Application access (Anyone)

Connected developer tools

Connect ˅

Development information will be displayed from these connected applications. Learn more.

Name	Application	Application URL	Capabilities
Bitbucket Cloud	Bitbucket Cloud	https://bitbucket.org	Create branches
			Create and view pull requests
			View commits
			View builds

Bamboo

Bitbucket Cloud

Bitbucket Server

FishEye / Crucible

GitHub Enterprise

Figure 1.28 – Connect Bitbucket Cloud

4. In the **Jira requests access** page, authorize Jira to access the Bitbucket workspace where you
    created the repo by selecting the workspace and clicking **Grant Access**.

Jira requests access

This app is hosted at https://wengroupllc.atlassian.net

Read and modify your account information

Read your repositories and their pull requests

Administer your repositories

Authorize for workspace

wengroupllc

Allow Jira to do this?

Figure 1.29 – Grant Bitbucket workspace access to Jira

5.   On the **DVCS accounts** page, select the repository or repositories to allow **Smart Commits**, and then click the **Link Bitbucket Cloud workspace** button.

DVCS accounts

After you connect your Bitbucket Cloud workspace to Jira and include issue keys in branch names, commit messages, and pull requests for linking repositories to your projects, you will be able to:

- Create branches and pull requests from your Jira project
- View development information within your linked Jira issue

Both GitHub cloud and enterprise accounts no longer use the DVCS connection, read more details here.

[Link Bitbucket Cloud workspace]

wengroupllc   Visit workspace •••

Repository	Smart Commits
Component1	☐
DevOps with Atlassian Chapter files	☐
DevOps1	☐
first	☐
repo2	☐
second	☑

Figure 1.30 – Add Repository and Link Bitbucket Cloud workspace

By connecting a Bitbucket repository to a Jira project, as you did by following the preceding instructions, you connect the reason for doing the work (the why) with the implementation of the work (the how).

2

Discovering Customer Needs with Jira Product Discovery

We start our exploration of the **DevOps** life cycle by looking at the typical trigger for a value stream, an important construct for DevOps thinking. This trigger is typically an idea or new thought that can be expressed as a new product or feature. **Jira Product Discovery** (**JPD**), a product in the Atlassian Jira family, allows users to create and track work items called **ideas** to record the ideation process. After a decision has been made to proceed into implementation, it can proceed into development as an **epic** in Jira.

In this chapter, we look at using JPD for storing, elaborating, and prioritizing ideas. The use and maintenance of ideas allow for better organization and easy determination of which ideas should follow into implementation. Because JPD, like the other Atlassian tools we talk about in this book, is based in the Atlassian cloud, we limit our discussion to the installation and configuration of the Atlassian cloud. As such, the recipes in this chapter include the following:

- Adopting JPD
- Creating a JPD project
- Viewing ideas
- Creating ideas
- Adding supplementary information in fields, comments, attachments, and insights
- Delivering ideas for development in Jira

Let's now start by installing JPD onto our Atlassian cloud site.

Adopting JPD

JPD can be installed using one of the following two licensing plans:

- A Free plan, for up to three licensed users (called creators)

- A paid Standard plan for larger teams, that have more than three creators

This recipe will focus on installing the Free plan. Organization admins can switch plans from Free to Standard or vice versa when the need arises.

Getting ready

As we saw in *Chapter 1*, there are different levels of administration on the Atlassian cloud. Some functions can only be performed by the organization admin, while others may be shared between the organization admin and the product admin (in this case, a Jira admin).

This recipe can only be performed by the organization admin.

How to do it...

Follow these instructions to install JPD onto your Atlassian cloud site as discussed in the introduction:

1. As the organization admin, go to the grid icon on the top-left corner of any Atlassian cloud product page to switch products on the site you want JPD installed. If you are responsible for multiple Atlassian cloud sites, find the site you want to administer.

2. Sites with a JPD entry in the **RECOMMENDED FOR YOUR TEAM** section do not have JPD installed. This can be seen in the following illustration:

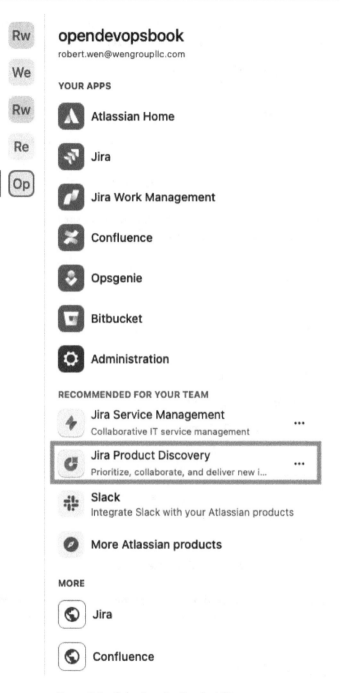

Figure 2.1 – Selecting Jira Product Discovery

3.   Selecting either entry of JPD will take you to the **Jira Product Discovery** product page, which describes the features and benefits of JPD. Select the **Try it now** button.

Jira Product Discovery

Move from product discovery to product delivery - all in Jira

Give product teams a dedicated tool to capture and prioritize ideas, connect business and tech teams, and align everyone.

Figure 2.2 – Opting in JPD

4.   Atlassian will then add the Free plan to your Atlassian cloud site in your organization. When complete, you will see the following message:

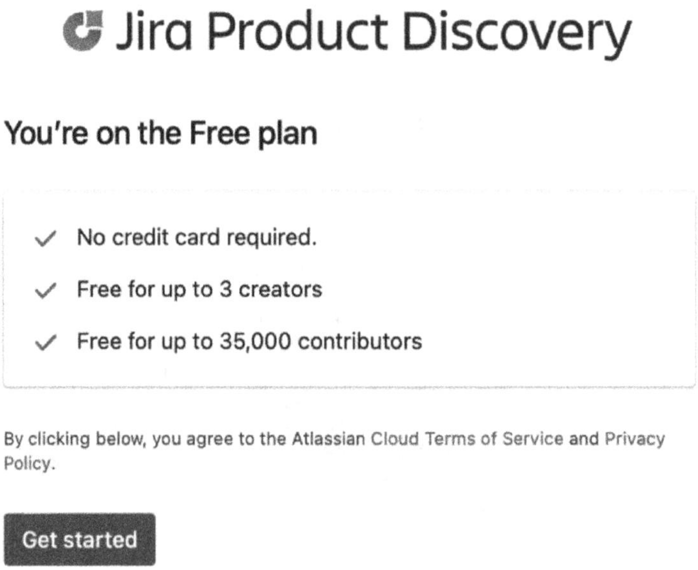

Jira Product Discovery

You're on the Free plan

✓   No credit card required.

✓   Free for up to 3 creators

✓   Free for up to 35,000 contributors

By clicking below, you agree to the Atlassian Cloud Terms of Service and Privacy Policy.

Get started

Figure 2.3 – JPD success message

After adding the product, the organization admin will be prompted to create a JPD project. In addition, a Jira admin may need to create a new JPD project to store new product ideas. Let's examine both use cases in the next recipe.

Creating a JPD project

As with other Jira products, such as Jira and Jira Service Management, JPD uses the notion of projects to store work and manage the people who perform the work.

JPD projects will store the work in terms of ideas, which is a potential starting point for a new product or product feature. People who work on ideas can either be creators, licensed users who create the ideas, or non-licensed contributors, who add supplementary data and sentiment as stakeholders, who can decide whether the idea is suitable for development.

How to do it...

One situation where a JPD project is created occurs when the JPD product is added by the organization admin. Let's now follow up on what we saw in the previous recipe.

Adding a JPD project after product addition

Jira Product Discovery prompts for the creation of an initial project after it is successfully installed onto an Atlassian cloud site. Let's view the steps involved in creating the project:

1. Once JPD has been successfully installed on the Atlassian cloud site, the organization admin can click the **Get started** button on the initial **Jira Product Discovery** page, where they will see a window that will prompt them to create the first project by adding a project name and key.

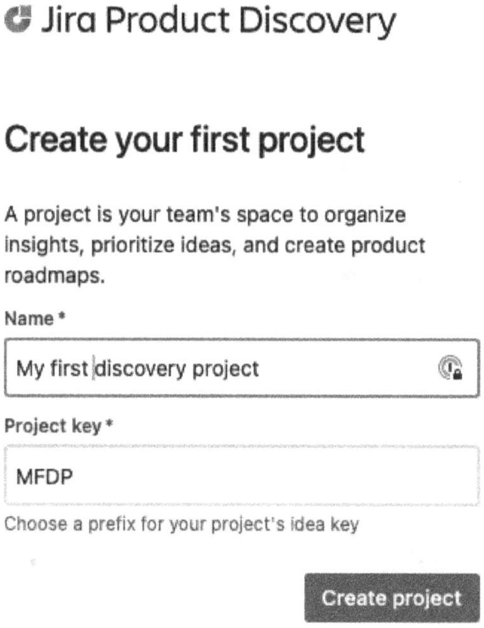

Figure 2.4 – The Create your first project window

2. After entering the required name and key and clicking **Create project**, the JPD project is successfully created. Project names and keys must be unique. Jira will display an error and not allow project creation if the name or key is identical to an existing project.

Creating a new JPD project

A Jira admin can create a JPD project through the same user interface that is used for Jira and Jira Service Management. To create a new JPD project from the Jira interface, execute the following steps:

1. From the **Projects** dropdown on the menu bar, select **Create project**, as seen in the following screenshot:

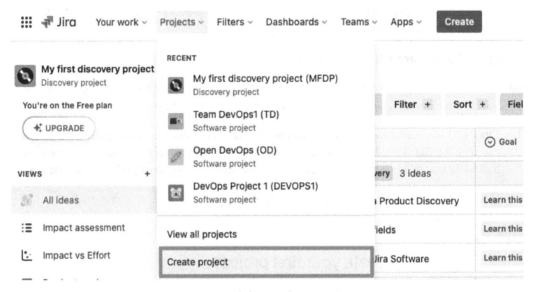

Figure 2.5 – Selecting Create project

2. Select the project template on the **Project templates** page. The JPD project template is located in the **Product management** category.

Figure 2.6 – Select a project template

3. On the following page, select **Use template**. This sets up the project to have the correct issue type and views that JPD uses.

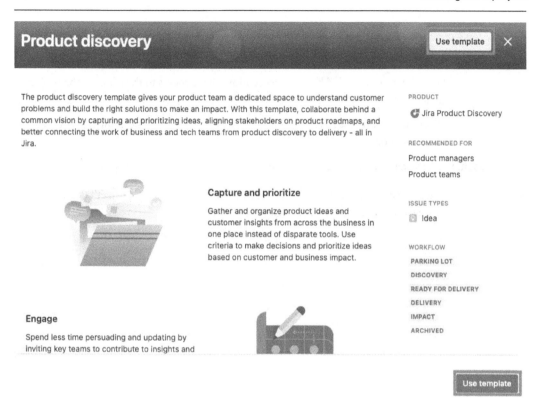

Figure 2.7 – Selecting Use Template

4. On the **Add project details** page, add the project details. Give your project a meaningful name and key. Select the level of access and press **Create project**, as illustrated here.

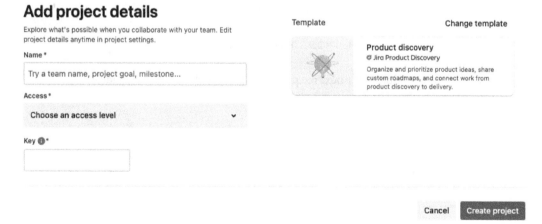

Figure 2.8 – Create project

5. The level of access selected has implications with who counts as a creator and thus is billable. There are three levels of access. These levels are defined as follows:

    •   **Private**: Only the users specified by the project admin have access to the project. The project admin can also determine who is a creator and who is a contributor.

    •   **Limited**: Everyone has access to the project. The project admin has the ability to determine who can act as creator.

    •   **Open**: Everyone has access to the project. Anyone with product access is given the creator role.

6. Once the project has been successfully created, any member will see the project's default view of ideas. This view may resemble the following screenshot:

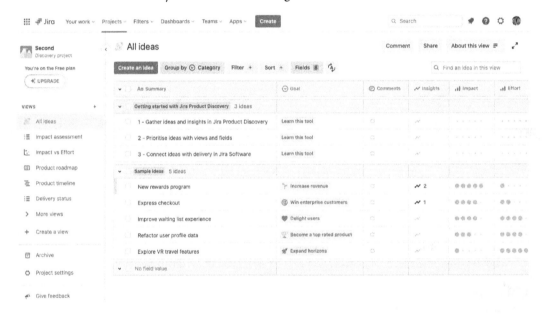

Figure 2.9 – JPD project view

From here, we can start looking at our next recipe: taking a deeper look at viewing ideas.

Viewing ideas

ideas are the key items stored in JPD projects. Ideas are displayed and organized using **views**.

Views provide several important functions. They can be used to organize and order ideas in a project based upon criteria such as when they could be completed on a potential roadmap or what teams could perform the work, and they can be used to compare competing ideas based on impact.

Views are based on the following four view types, which allow for different ways of organizing ideas:

- **List**: This presents ideas in a table format with relevant fields. Default examples include impact assessment and delivery status.
- **Board**: This shows ideas against a board with columns representing statuses or dispositions. A default example is the product roadmap.
- **Matrix**: This is used to compare ideas based on two criteria. An example of this is impact versus effort.
- **Timeline**: This shows ideas organized against when development and release will occur. A default example is the product timeline.

Let's take a closer look at working with views.

How to do it...

Views are capable of the following functions:

- Comments
- Grouping ideas
- Filtering ideas
- Sorting ideas
- Editing idea fields
- Creating ideas

Direct work on creating and editing ideas instead of from within a view will be discussed in the next section. For now, let's take a look at how to organize our ideas inside of views.

Adding comments

Anybody, creator and contributor alike, with access to the JPD project can comment on any view.

Let's look at adding comments on our ideas from the view:

1. To create a comment on a view, select the **Comment** button, as depicted on the following view:

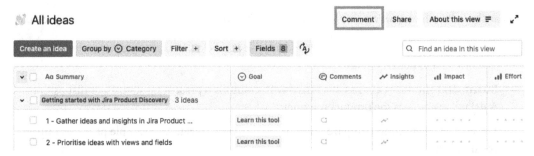

Figure 2.10 – Comment button for the view

2. A window will open below the **Comment** button, as shown in the following image. Click inside the text box to begin creating your comment.

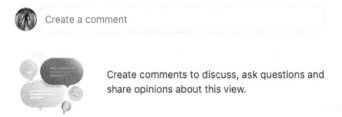

Figure 2.11 – Starting to create a comment

3. The text box will expand, showing controls for comment styling and content. Add your comment and click **Create** to finish creating your comment.

Figure 2.12 – Comment Creation box

4. Once a comment is created, the **Comment** button will indicate the number of comments for that view. Existing comments will appear below the **Create a comment** text area.

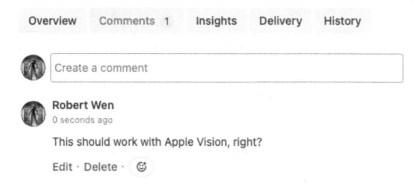

Figure 2.13 – Existing comments and number of comments

Grouping ideas

You can group ideas based on common criteria such as customer segment, creators, goals, or other qualities that help determine which ideas should be prioritized higher. Grouping ideas is available in the list, board, and timeline view types. Follow these steps to group ideas in a view:

1. To group ideas in a view, select the **Group by** button in the view, as depicted here.

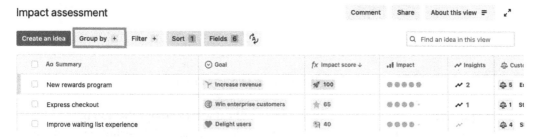

Figure 2.14 – Creating a grouping

2. The criteria to select the grouping options opens from the right, as shown in the following snippet. Select which idea field to use to group ideas from the pulldown menu.

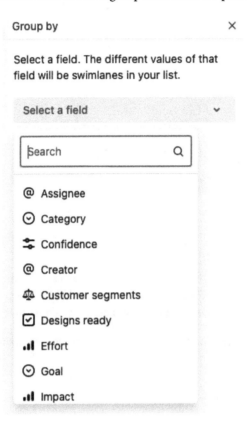

Figure 2.15 – The Group by selection menu

3. Once selected, the view will change, and ideas will fall within groups that can be expanded to see the ideas inside. The categorization will divide up the issues into swimlanes within the view, as seen in the following screenshot:

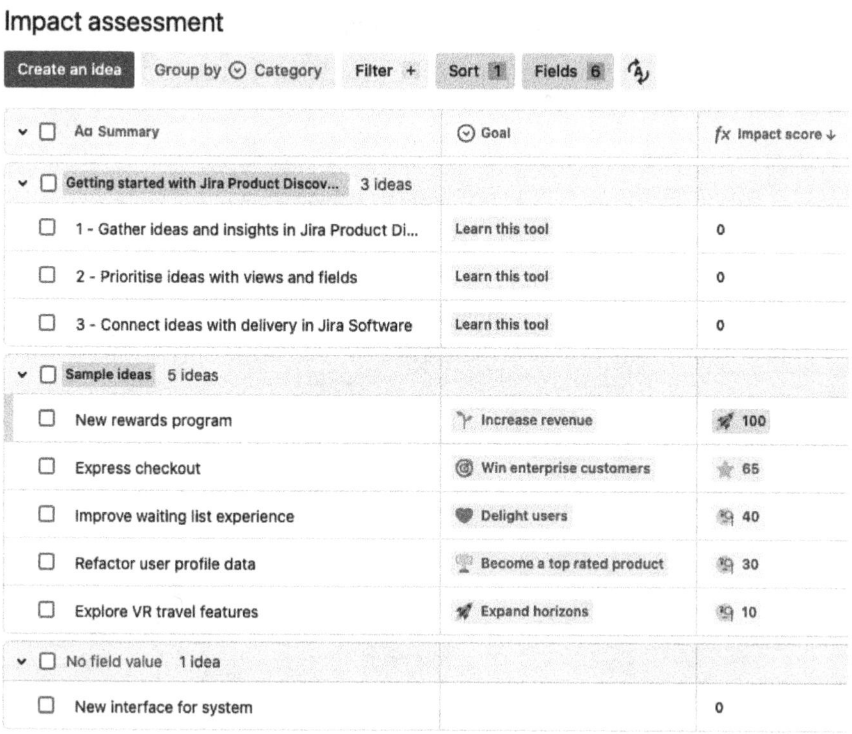

Impact assessment

	Aα Summary		⊙ Goal		ƒx Impact score ↓
⌄ ☐	**Getting started with Jira Product Discov...** 3 ideas				
	☐ 1 - Gather ideas and insights in Jira Product Di...		Learn this tool		0
	☐ 2 - Prioritise ideas with views and fields		Learn this tool		0
	☐ 3 - Connect ideas with delivery in Jira Software		Learn this tool		0
⌄ ☐	**Sample ideas** 5 ideas				
	☐ New rewards program		⌄ Increase revenue		🚀 100
	☐ Express checkout		◎ Win enterprise customers		⭐ 65
	☐ Improve waiting list experience		♥ Delight users		🦊 40
	☐ Refactor user profile data		🏆 Become a top rated product		🦊 30
	☐ Explore VR travel features		🚀 Expand horizons		🦊 10
⌄ ☐	**No field value** 1 idea				
	☐ New interface for system				0

Figure 2.16 – View divided by groups

Filtering ideas

All view types can be filtered to allow ideas matching the filter criteria to appear on the view. This may be done to focus on specific ideas as an aid for prioritization. Execute the following steps to set up filters for your ideas:

1. To set a filter, press the **Filter** button.

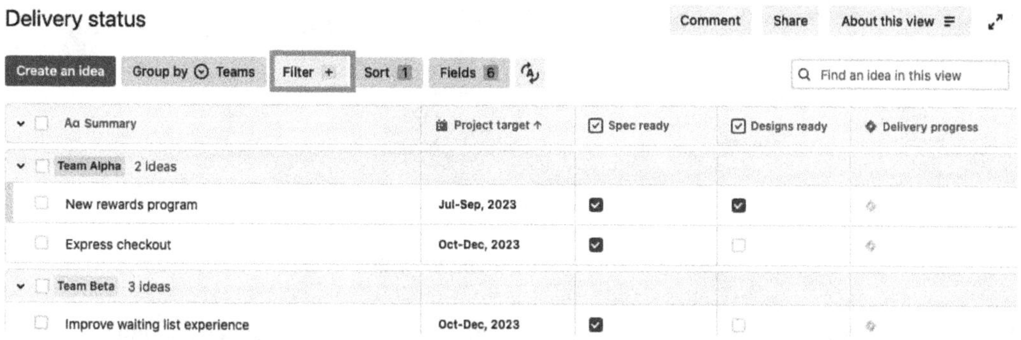

Delivery status

	Aα Summary		🗓 Project target ↑	☑ Spec ready	☑ Designs ready	◆ Delivery progress
⌄ ☐	**Team Alpha** 2 ideas					
	☐ New rewards program		Jul-Sep, 2023	☑	☑	◌
	☐ Express checkout		Oct-Dec, 2023	☑	☐	◌
⌄ ☐	**Team Beta** 3 ideas					
	☐ Improve waiting list experience		Oct-Dec, 2023	☑	☐	◌

Figure 2.17 – Creating a view filter

2. A **Filters** area opens from the right. Select **Add a filter** to create your filter.

3. As shown here, a pulldown with idea fields will appear. Select the field and desired field options to use as filter criteria.

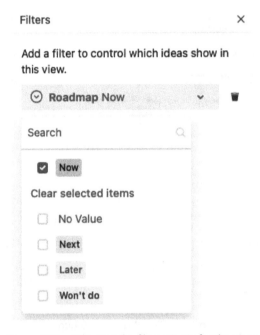

Figure 2.18 – Selecting the filter criteria for the view

4. Click the **X** in the top-right corner when finished with the filter selection.

By following the preceding steps, you can choose to see what ideas are shown in your view through filters. Only ideas that meet the filter criteria will appear in the view, as seen in the following screenshot:

Figure 2.19 – A view with filters applied

Let's look at one last way of organizing ideas within a view. Sometimes, ideas can be put into a specific order. Sorting may help in determining that order.

Sorting ideas

You can determine the order in which ideas appear in a view by setting the sorting order. Establishing a sorting order helps with the organization and prioritization of ideas and may aid in determining which idea should move to implementation. This is available for list, board, and timeline view types only. Implement the following steps to sort your ideas:

1. To allow for sorting, press the **Sort** button.

Figure 2.20 – Pressing Sort

2. The **Sort** window will open from the right. Select the idea field or fields to use for your first and secondary sort, and a tertiary sort if needed.

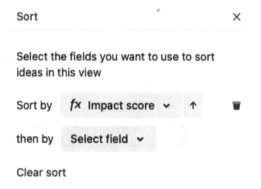

Figure 2.21 – Setting the sort order

3. Click the **X** in the top-right corner when finished with the sort selection.

If you look at your view, you should see that your ideas are sorted according to the criteria you set. The following screenshot shows the effect of sorting by ascending **Impact score**, which reverses the default sorting order:

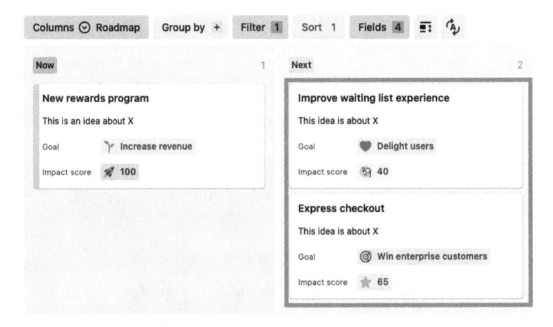

Figure 2.22 – View with a sort order set

At this point, we're ready to look at ideas in JPD. Let's take a look at the *Creating ideas* section.

Creating ideas

Ideas are the primary artifact in JPD. Ideas allow for descriptions and justifications for new products and features. The idea becomes the repository of all supplementary information through the use of the following features in ideas:

- Fields
- Comments
- Attachments
- Insights

Let's look at creating ideas by filling in the preceding features both on their own and from within the context of a view.

Getting ready

By default, the actual creation of ideas can only be done by those who have the creator role in a JPD project.

Jira admins can allow those with contributor access to create ideas in a project by selecting the **Project settings** option on the sidebar.

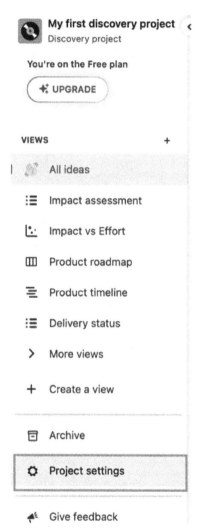

Figure 2.23 – Selecting Project settings

From the **Project settings** sidebar options, expand on **Features** and select **Create ideas**. The page should resemble the following screenshot:

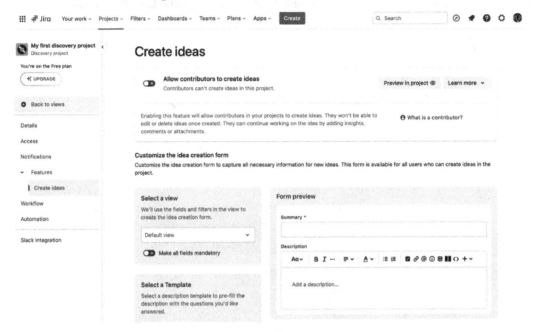

Figure 2.24 – The Create ideas settings

Switch the toggle on **Allow contributors to create ideas** so that it is enabled.

Note that while enabled, contributors can create ideas but cannot edit them or delete them. On any idea, contributors can add comments, attachments, and insights.

How to do it...

In a JPD project, there are multiple ways of creating new ideas. The typical creation places include the following:

- The blue **Create** button on the navigation bar
- In any view from either the **Create an idea** button or the **Add idea** control

Let's look at how to create ideas using these modes.

Creating an idea from the Create button

Perform the following actions to create an idea using the **Create** button:

1. Once you have determined who can create an idea, they can press the blue **Create** button on the navigation bar.

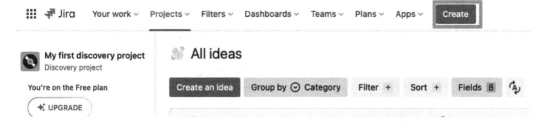

Figure 2.25 – Using the Create Button

2. A window appears with the standard fields needed for the idea. Fill in those fields and click the **Create** button.

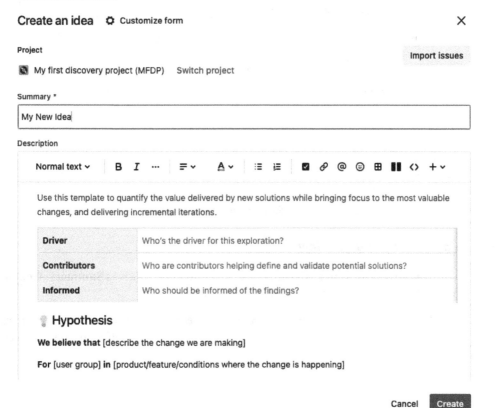

Figure 2.26 – The Create an idea window

3.   The Jira admin can change what fields appear on the window by creating forms. This may also include pre-filling the **Description** field by using a template.

Following the preceding instructions creates an idea that can be seen in any view of the JPD project. An example of such a created idea is illustrated in the following screenshot.

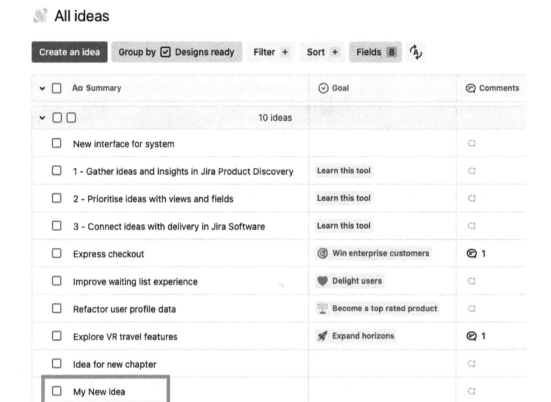

Figure 2.27 – A view with a newly-created idea

Creating an idea from a list view

A list view in JPD displays ideas in a tabular format. Common fields such as **Summary**, **Goal**, **fx Impact score**, and **Effort** appear as columns in the table.

Default views such as **All Views**, **Impact assessment**, and **Delivery status** are set up as list views. These list views can be accessed from the **Views** section in the sidebar. In addition, you can create your own custom list views.

Let's take a look at how to create a new idea when you are in a list view:

1. With the view visible, select the **Create an idea** button, as depicted in the following illustration:

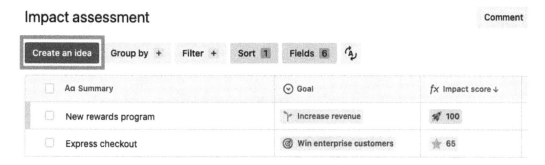

Figure 2.28 – The Create an idea button

2. A row opens at the top, allowing you to enter the summary of the new idea. Type your summary and press *Enter*.

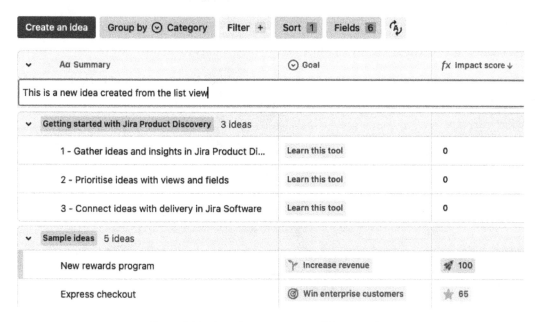

Figure 2.29 – Entering a summary of the new idea

The new idea should now be visible in your list view.

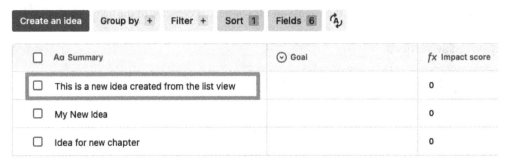

Figure 2.30 – New idea in the list view

Let's now learn more about roadmap views and how to create new ideas in that view.

Creating an idea in a roadmap view

A roadmap view displays ideas in terms of when they can be implemented. In a roadmap view, columns representing **Now**, **Next**, and **Later** points of time hold cards that represent ideas. This allows for the visualization of which ideas should be done sooner rather than later.

Roadmap views are accessed by selecting them in the **Views** section of the sidebar.

Let's examine how to create a new idea when we're in the **Product roadmap** view:

1. On the roadmap, select the column where you'd like the idea to appear and press the + **Add idea** button at the bottom of the column.

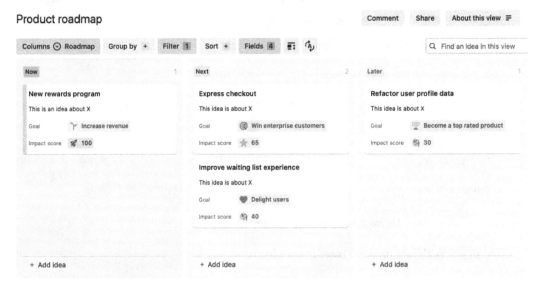

Figure 2.31 – Add an idea to the roadmap view

2. In the search area that appears, type in the summary for a new idea.

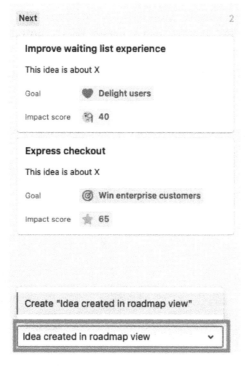

Figure 2.32– Entering a summary for a new idea

We now have a new idea that is on the roadmap.

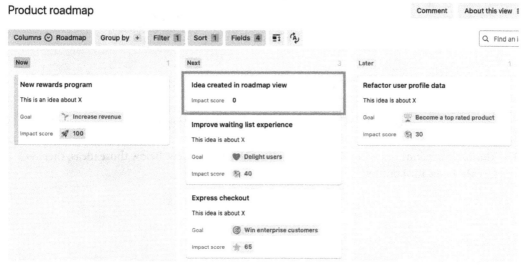

Figure 2.33 – A new idea in the roadmap view

Creating an idea in a matrix or timeline view

Matrix views and **timeline** views display ideas in two-dimensional space where the idea's position represents its importance in a matrix view, such as impact versus effort or when it is planned to be implemented in a timeline view.

Like the other views, matrix and timeline views can be accessed from the **Views** section in the sidebar.

Perform the following steps to create a new idea when you are in a matrix or a product timeline view:

1. On a matrix view or a product timeline view, a blue circle with a plus sign will appear in the lower right corner. Select the blue circle to create a new idea.

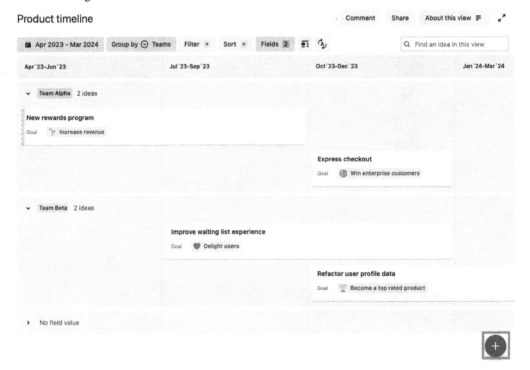

Figure 2.34 – Create an idea in the product timeline view

2. The circle expands, showing ideas that can be added to the view. Below those ideas, there is a **Create a new idea** button.

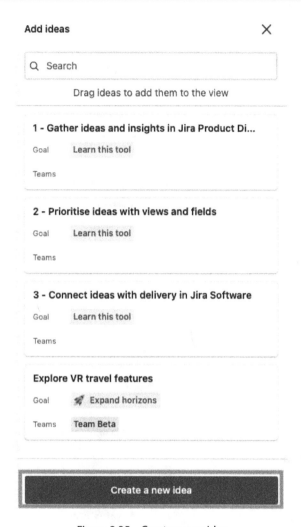

Figure 2.35 – Create a new idea

3. The button changes to a text field, as seen here, allowing you to enter the summary for a new idea. Complete the creation by pressing *Enter*.

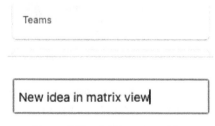

Figure 2.36 – Adding a new idea

The new idea can then be placed where it's needed on the product timeline or matrix view by dragging from the **Add ideas** area onto the view.

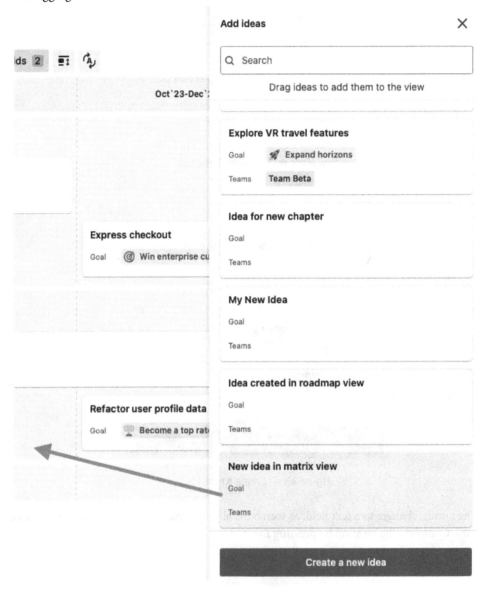

Figure 2.37 – Dragging a new idea onto the timeline view

Once we have an idea created, we may want to add supplementary information with it to record learning and help with decision making. The next recipe will demonstrate how to add additional information to ideas.

Adding information to ideas

Once an idea is created, both creators and collaborators may add context to it in order to show the evolution of the idea and help facilitate decision making. The common ways of adding this information include using the following:

- Fields
- Comments
- Attachments
- Insights

Let's examine the common methods used to add this information.

Getting ready

Permissions can dictate who is allowed to add information. Fields can only be modified by creators. Both creators and collaborators can add comments, attachments, and insights.

How to do it...

Fields, comments, attachments, and insights can be added from the following locations:

- On the page that appears when selecting an idea
- From an expanded right panel that appears when clicking on an idea in a view
- Directly in the view (for list view types)

Let's look at how to add this information.

Adding on the idea page

Let's look at adding data for the idea on its own page:

1. Ideas can be displayed on their own page often by clicking on the ID of the idea from anywhere in the project, but most commonly from a view. An idea page is shown with the locations for fields, comments, attachments, and insights labeled.

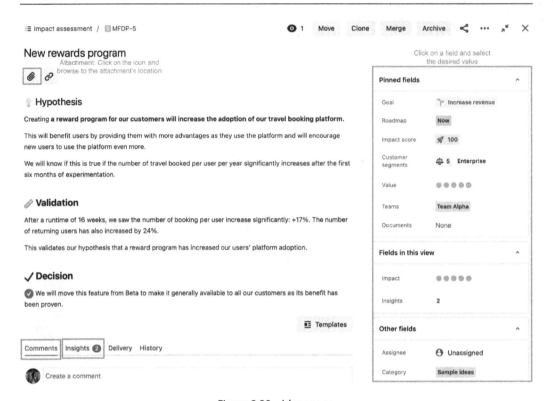

Figure 2.38 – Ideas page

2.   **Comments** and **Insights** are set up as tabbed fields. Select **Comments** to view all comments and enter new ones. Enter a new comment in the text area and finish by clicking the **Create** button.

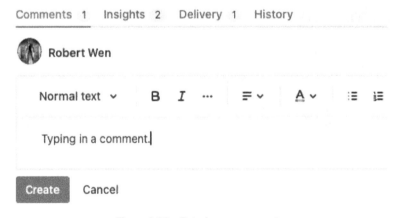

Figure 2.39 – Entering a comment

The comment is seen with other comments for that idea.

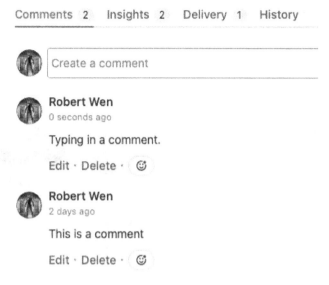

Figure 2.40 – A new comment with existing comments

3. Select **Insights** to view links to other sources or create such a link yourself by clicking the **Create an insight** button. Enter the link that is the source of the insight and enter descriptive text in the text area. Click the **Create** button when complete.

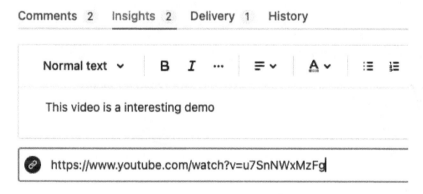

Figure 2.41 – Entering an Insight

The new insight is then at the top of the list of insights.

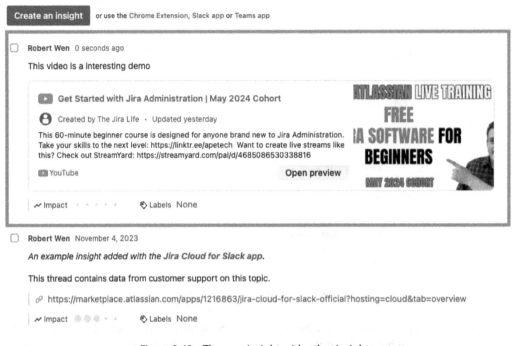

Figure 2.42 – The new insight with other insights

We have seen that we can add relevant information about an idea on its own page. Let's look at doing the same things from the idea panel that opens when selecting an individual idea in a view.

Adding from the idea panel

You can click on any idea in a view to allow the idea panel to display on the right of the view. The idea panel is depicted here.

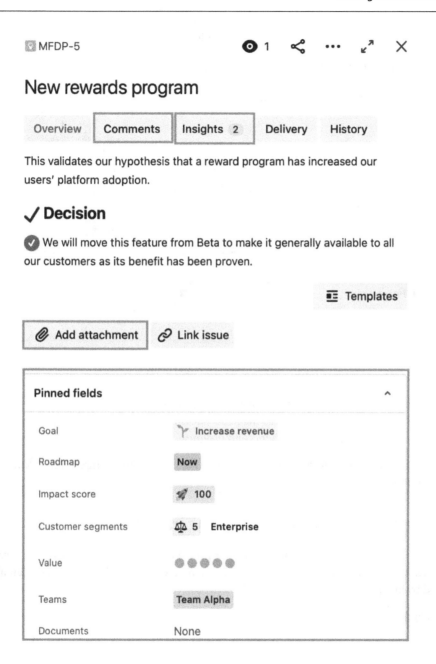

Figure 2.43 – Idea panel

Comments and **Insights** can be seen and added from their own buttons. Attachments can be added from the **Add attachment** button. Fields are modified by scrolling to the field, clicking on it, and selecting the desired value.

The preceding steps have allowed us to add relevant information to an idea using the idea panel.

Modifying from a list view

A list view is laid out as a table, as seen in the screenshot. Rows represent ideas, with columns representing fields.

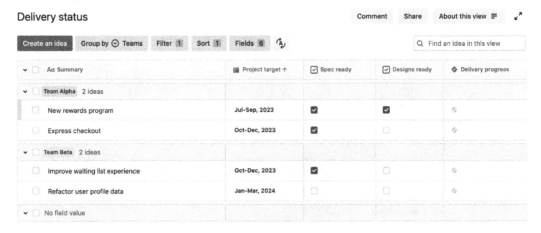

Figure 2.44 – A list view and fields

Creators are able to modify an idea's fields by going to the intersection of the row, which represents the idea, and the column, which represents the field. Clicking on the cell allows for the modification of the field. Some features, such as comments and insights, can be set up as additional columns.

Once an idea has been approved to implement, we need to link it with an implementation. The next recipe will allow that by creating a linked epic in Jira.

Delivering ideas for development in Jira

When ideas are ready for development, they can be linked to epics in Jira. As the epic is created, a link is established between the original idea and the generated epic. Delivery progress can also be determined by looking at the status of linked Jira issues.

Let's take a look at linking JPD with Jira.

Getting ready

Before creating the epic, please verify the following:

- There is at least one Jira project available to contain the epic.
- The person creating the epic has the appropriate permissions to create the epic in the target project. This same person should be a member of the JPD project or a Jira admin.

When both conditions are met, it becomes easy to create the epic as a delivery ticket in JPD.

How to do it...

Creating an epic as a delivery ticket can be done from the following two places:

- On the idea's page
- On the idea panel in a view

Let's examine how to create the epic from these locations.

Creating the epic from the idea page

To create an epic in Jira that is linked to the idea in JPD, perform the following steps:

1. On the idea page, click on the **Delivery** tab.

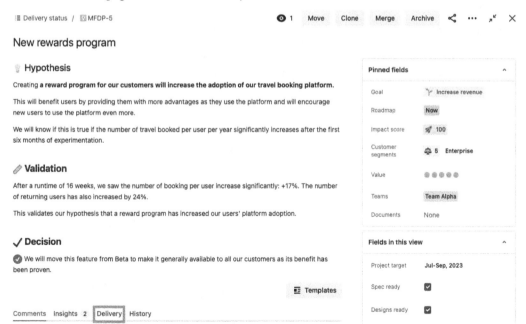

Figure 2.45 – Selecting the Delivery tab on the idea page

2. The **Delivery** tab will expand, giving you the following two options:

    *   Creating a new delivery ticket (epic)

    *   Creating a link to an existing delivery ticket

3. Selecting **Create a delivery ticket** displays text areas for the project, issue type, and summary, as seen here.

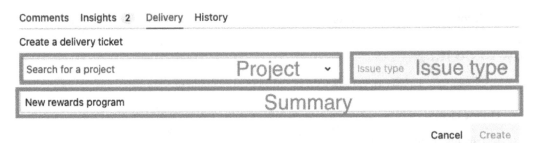

Figure 2.46 – Create a delivery ticket

4. Select the project and issue type, and complete the desired summary before clicking **Create**.

An epic is now present in the Jira project and is linked to your idea.

Creating the epic from the idea panel

Let's look at creating an epic and linking it to our idea from the idea panel:

1. From the view, select an idea to display the idea panel on the right. You should see the **Delivery** tab near the top of the panel, as depicted here:

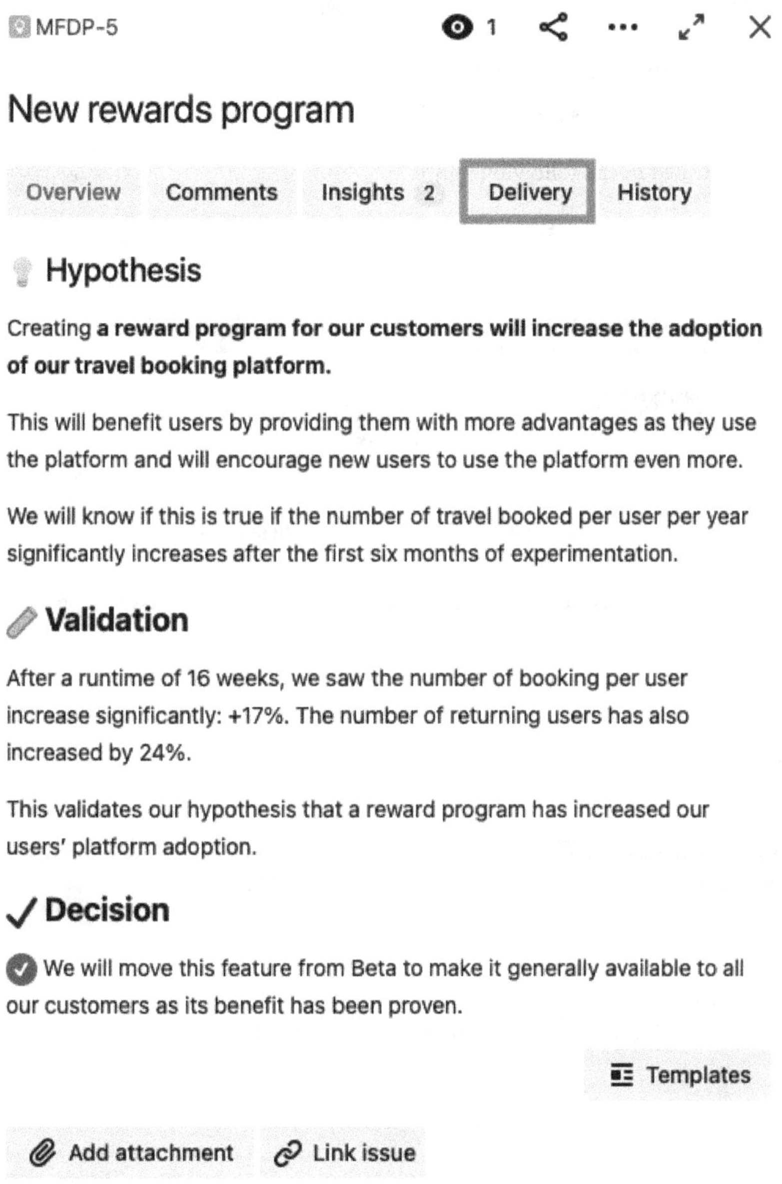

MFDP-5
1
New rewards program
Overview
Comments
Insights 2
Delivery
History
Hypothesis
Creating a reward program for our customers will increase the adoption of our travel booking platform.
This will benefit users by providing them with more advantages as they use the platform and will encourage new users to use the platform even more.
We will know if this is true if the number of travel booked per user per year significantly increases after the first six months of experimentation.
Validation
After a runtime of 16 weeks, we saw the number of booking per user increase significantly: +17%. The number of returning users has also increased by 24%.
This validates our hypothesis that a reward program has increased our users' platform adoption.
Decision
We will move this feature from Beta to make it generally available to all our customers as its benefit has been proven.
Templates
Add attachment
Link issue

Figure 2.47 – The Delivery tab in the idea panel

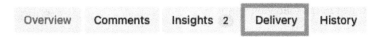

2. The **Delivery** tab will expand, giving you the following two options:

  - Creating a new delivery ticket (epic)

  - Creating a link to an existing delivery ticket

3. These are illustrated in the following figure.

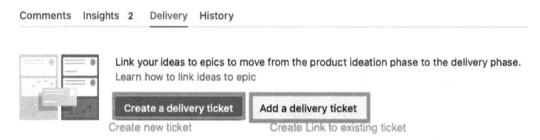

Figure 2.48 – Create a delivery ticket

4. Selecting **Create a delivery ticket** displays text areas for the project, issue type, and summary, as seen here.

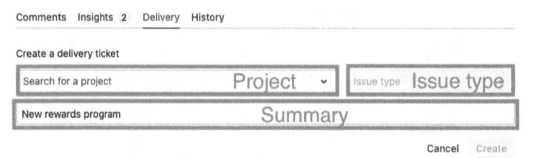

Figure 2.49 – Populating delivery ticket information

The epic appears in the list of delivery tickets.

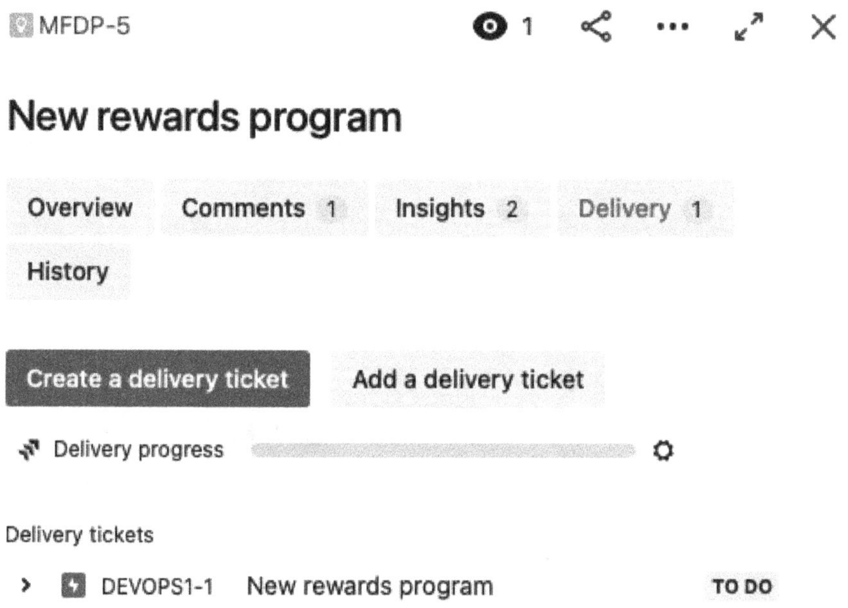

Figure 2.50 – Display of delivery ticket

In this chapter, we have explored the capabilities of JPD that allow us to track and prioritize product ideas and make decisions about when they should be sent to development teams for implementation. This new work starts the DevOps cycle that Development and Operations teams use to develop, release, and maintain products.

Planning and Documentation with Confluence

As we begin the development process, we must look for a convenient place to document our progress, specify what the scope of work is, and state how we'll go about the development process. For many organizations, that place is **Confluence**.

Integrating Confluence with other products in the DevOps toolchain allows you to link artifacts from those other tools into Confluence documentation pages. The contents of these pages may include plans and other project data, as well as code contents from Bitbucket. By allowing this tight integration between Confluence pages and artifacts in Jira and Bitbucket, the documentation artifacts that are created as pages in Confluence have a richer context, enabling richer meaning and better understanding.

In this chapter, we will leverage integrations between Confluence and the other Atlassian DevOps toolchain tools that are available. To do so, we'll cover the following recipes:

- Creating Confluence pages linked to Jira projects
- Displaying Jira issues from Confluence pages
- Viewing reports generated in Jira on Confluence pages
- Viewing Jira roadmaps in Confluence pages
- Viewing linked pages from other applications using Smart Links

Technical requirements

To complete the recipes in this chapter, you'll need to have created an Atlassian Cloud site with both Jira and Confluence installed. Connecting Confluence to Jira was covered in the *Connecting Confluence* recipe in *Chapter 1*.

In addition, a Jira administrator and possibly a Jira project administrator must verify you have the Browse Projects permission enabled for the project you want to work with. A Confluence Space administrator must verify that you are in the correct group and have enabled **Pages, whiteboard, and Smart Links**, as illustrated in the following screenshot:

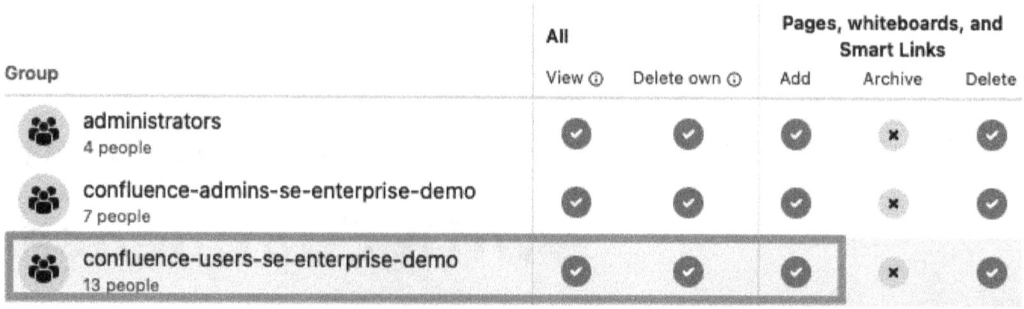

Group	All		Pages, whiteboards, and Smart Links		
	View ⓘ	Delete own ⓘ	Add	Archive	Delete
👥 administrators 4 people	✓	✓	✓	✗	✓
👥 confluence-admins-se-enterprise-demo 7 people	✓	✓	✓	✗	✓
👥 confluence-users-se-enterprise-demo 13 people	✓	✓	✓	✗	✓

Figure 3.1 – Space permission to add Pages, whiteboard, and Smart Links

Now that we've covered the prerequisites, let's start by linking Confluence to a Jira project.

Creating Confluence pages linked to Jira projects

We learned how to integrate Jira and Confluence in *Chapter 1*. Within every Jira project, members can create and modify Confluence content directly from Jira.

The integration between Confluence pages and Jira projects allows you to keep documentation in close context with the events and artifacts that are produced by the project. Here are some examples of documentation:

- User documentation
- Requirement specifications
- Meeting notes
- Release notes

Let's learn how to create such content and connect it to a Jira project.

How to do it...

You can create Confluence pages without leaving Jira. This allows for greater productivity as it eliminates unnecessary context switching.

Let's see how that's done:

1. From your Jira project, select **Project pages** from the sidebar on the left, as shown in the following screenshot:

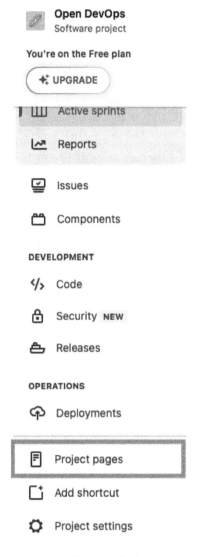

Figure 3.2 – Selecting Project pages

2.   The page will show any content from Confluence, such as pages or Spaces (collections of pages) that have already been connected. From here, you have the following options:

- Edit a currently connected page

- Create a new page at the top level

- Create a new child page of an existing page

Let's look at these options in more depth.

Editing an existing page

Follow these steps:

1.   On the **Project pages** page, find the page you wish to edit. Hover over the row corresponding to that page; you should see a pencil icon, as shown here:

Figure 3.3 – Selecting a page to edit

2.   Selecting the pencil icon opens a modal where the page is ready to edit. Make your changes and click **Update** to publish the changed page.

Here, you went to Confluence from Jira and edited the page via Confluence. The row updates to indicate that it was recently edited.

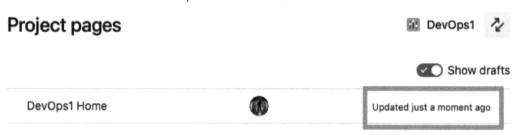

Figure 3.4 – Updated page in Jira

Creating a new page

Follow these steps:

1. To the right of the **Project pages** page is a section that shows relevant page templates.

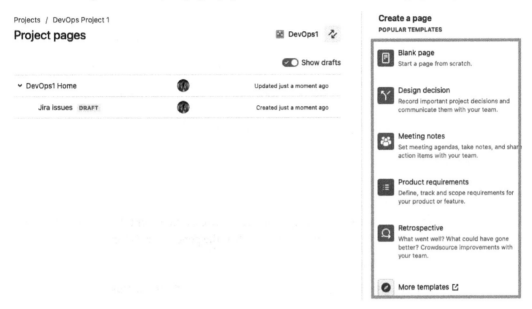

Figure 3.5 – Selecting a template for a new page

2. Selecting a template opens a new page where that template is being used in a modal, as shown here:

Figure 3.6 – Editing a new page

3. Make the desired changes and click **Publish** to make the page available to others in Confluence or Jira. The page will appear at the top level of the Space.

You should now have a brand-new page in Confluence that's connected to your Jira project.

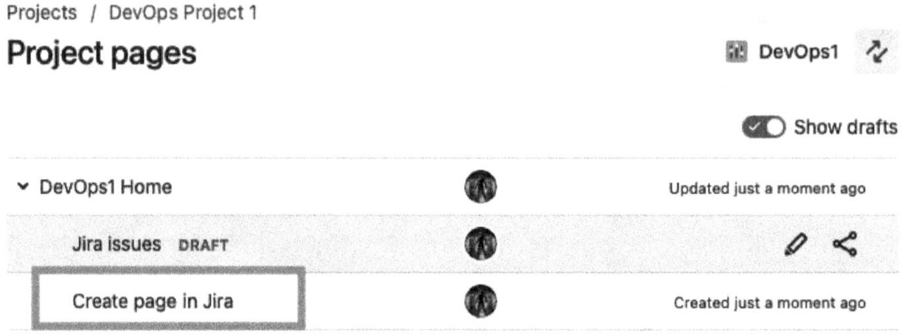

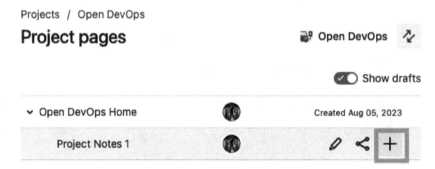

Figure 3.7 – New Page seen in Jira

Creating a child page

Follow these steps:

1. On the **Project pages** page, find the page you want to use as the parent. Hover over the row representing that page to view the various tools, such as the pencil icon. Select the plus sign (+).

Projects / Open DevOps

Project pages

Figure 3.8 – Selecting the parent page

2. A page you wish to edit will open as a modal. Make the desired changes and click **Public** to make the page available to others in Confluence or Jira.

So far, the content we've added to our Confluence pages has been text that we've edited using a default text editor. In the following recipes, we'll look at adding special content to these pages.

Displaying Jira issues on Confluence pages

Integrating Confluence with Jira allows for enhanced content to be added to Confluence pages for documentation purposes. The content that's added through integration is updated in real time, allowing the Confluence page to be dynamic.

You can integrate Jira issues into Confluence pages by using **macros**. These are add-on features that are native to Confluence or can be added through Atlassian Marketplace apps. They provide enhanced formatting and allow you to add content.

Let's take a closer look at how to use Confluence macros.

How to do it...

We will use macros to perform the following actions:

- Display existing Jira issues from the Jira project
- Create new Jira issues

Each action utilizes a different macro.

Displaying existing Jira issues

Let's look at the macro that displays existing Jira issues:

1. To view the available macros, select the plus sign (+) on the page you're editing or edit the modal in Jira, as shown here:

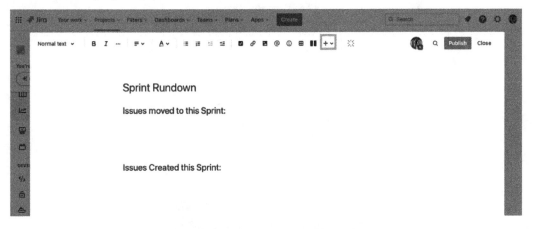

Figure 3.9 – Adding a macro

2. Search for the appropriate Jira-specific macro. In this case, we want to find the **Jira Issues** macro from the pulldown shown in the following screenshot. You can search by keyword or select a category. Select **View more** to view all categories.

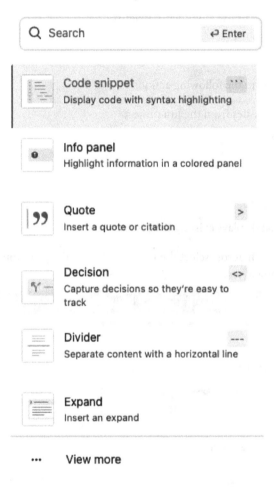

Figure 3.10 – Searching for a macro

3. When selecting **View more**, all the available macros can be accessed on a separate **Browse** screen by category or by searching for them. Select the macro you wish to use and click **Insert**

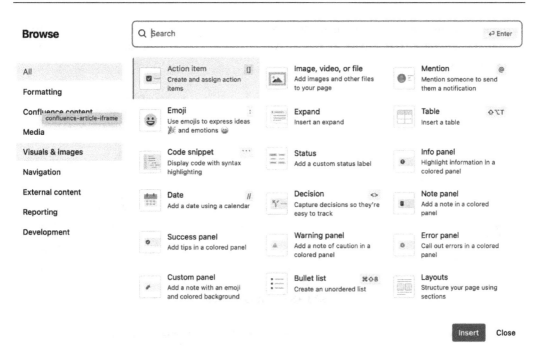

Figure 3.11 – Selecting a macro from the Browse screen

4. In our case, we want to find the **Jira Issues** macro. We can do this by searching for `jira` and selecting it from the possible options that appear.

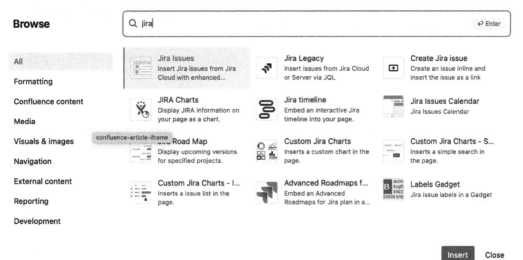

Figure 3.12 – Selecting the Jira Issues macro

5. The **Jira Issues** macro looks for issues to display through a **BASIC** search mechanism or a **Jira Query Language** (**JQL**) query for more advanced queries. Search for the desired issues and select **Insert issues**.

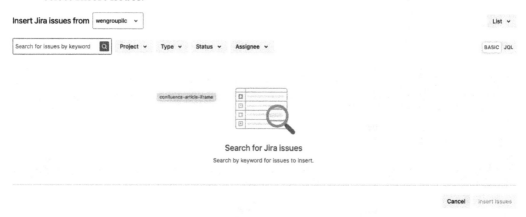

Figure 3.13 – Selecting Jira issues

Now that we know how to use macros to add content, let's look at using a macro to create new Jira issues.

Creating new Jira issues

To create a new Jira issue and link it to our Confluence page, we are going to use a different macro. Follow these steps to create new Jira issues through a macro:

1. We can look for Jira macros in the **Development** category, as shown in the following screenshot. Select the **Create Jira issue** macro and click **Insert**.

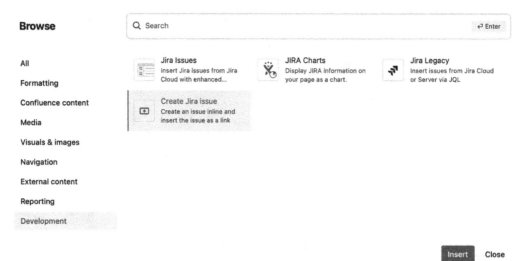

Figure 3.14 – Selecting the Create Jira issue macro

2. The fields that you can use to edit the new Jira issue will appear as a modal. Fill in the desired fields and click **Create** to record the new Jira issue.

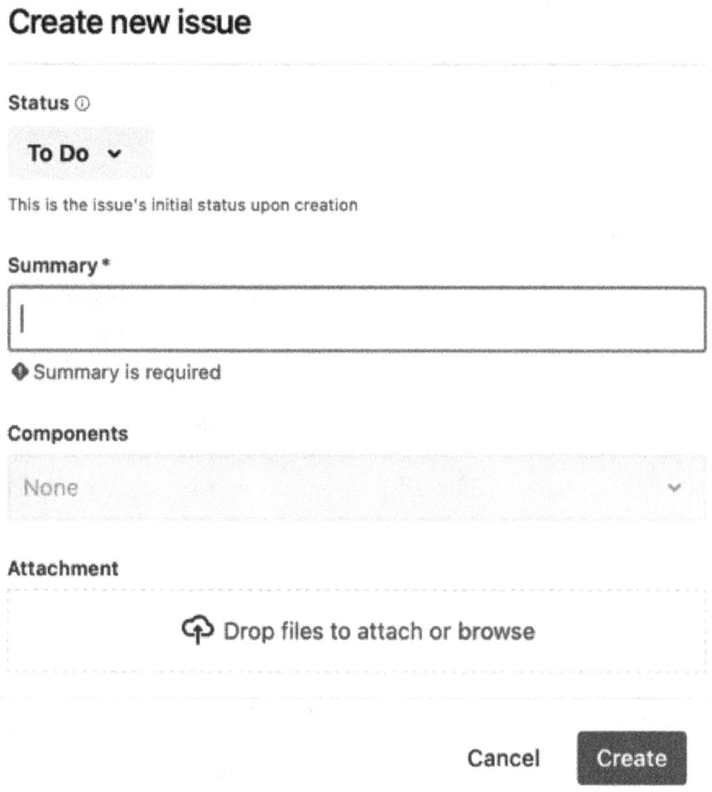

Create new issue

Status ⓘ

To Do ⌄

This is the issue's initial status upon creation

Summary *

◆ Summary is required

Components

None ⌄

Attachment

☁ Drop files to attach or browse

Cancel   Create

Figure 3.15 – Creating a Jira issue

Once created, the Jira issue can be formatted so that it can displayed on the Confluence page. You can choose from the following configurations:

- URL
- A single line of text
- As a card, with several fields showing
- Embedded onto the page

The desired formatting can be changed if the Confluence page is in edit mode.

With that, you should have a new Jira issue that's also present on the Confluence page.

See also

For more information about creating queries using JQL, check out the following resources:

- `https://www.youtube.com/watch?v=KC7vPPP2sQo`

- `https://www.youtube.com/watch?v=Bbvl9PqCePA`

- `https://www.atlassian.com/software/jira/guides/jql/overview#what-is-jql`

Viewing Jira reports on a Confluence page

The same reports that are created on Jira can be replicated and displayed on a Confluence page. Macros allow you to replicate and display these reports. These macros come built into Confluence or may be provided by third-party Atlassian Marketplace apps such as Custom Charts for Jira.

You might be wondering, "If the reports are available on Jira, why use Confluence?" The answer is that Confluence provides the perspective of viewing the reports within the greater context of the project. A report can be embedded within a Confluence page or as part of a Confluence Space, which keeps all of the project information together in one spot for easy lookup.

Let's examine how to enable these reporting macros on your project's Confluence page.

How to do it...

The following macros can be used for reporting in Confluence:

- **The JIRA report template**: This is a Confluence page template for text-based reports such as **Change log** or **Status report**.

- **The JIRA Charts macro**: This allows you to place custom 2D graphs, created versus resolved charts, or pie charts.

Let's learn how to create reports regarding Jira data in Confluence using these macros.

Using the JIRA Charts macro

We'll start by looking at the JIRA Charts macro:

1. The reporting macros can be found in the **Development** category when you're browsing for macros, as shown here:

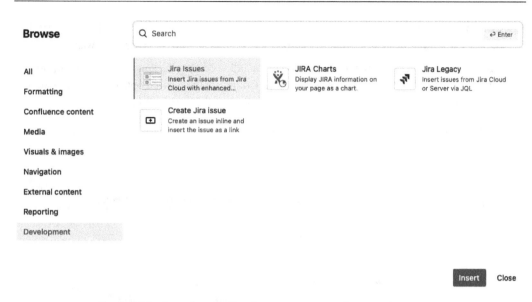

Figure 3.16 – Browsing the Development category for reporting macros

2. Select **JIRA Charts** and click **Insert**.

3. An alternate method for finding and selecting any Confluence macro can be done while editing the Confluence page. Just add a backslash character (/) to an empty line, as depicted here:

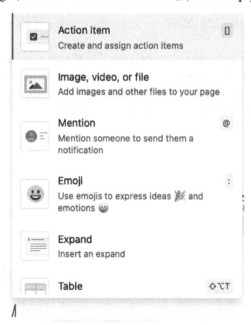

Figure 3.17 – Selecting a macro while editing a Confluence page

4. Continue typing the name of the macro (in this case, `JIRA Charts`). The built-in search function will limit the possible options as you type. When you find the macro you want, select it.

5. Once selected from either the browse modal or via inline editing, various configuration options for the reporting macro will appear in a modal. Configure the desired options, including data sources, refresh interval, and formatting. An example of this is shown here:

Figure 3.18 – Configuring the JIRA Charts macro

6. As we can see, the configuration includes selecting the type of chart (**Two Dimensional**, **Created vs Resolved**, or **Pie Chart**), specifying the data source through a JQL query or filter, and various **Display options**, including grouping the data. Once you've finished configuring the chart, click **Insert**. An example of doing such a configuration using the **Created vs Resolved** chart is shown in the following screenshot:

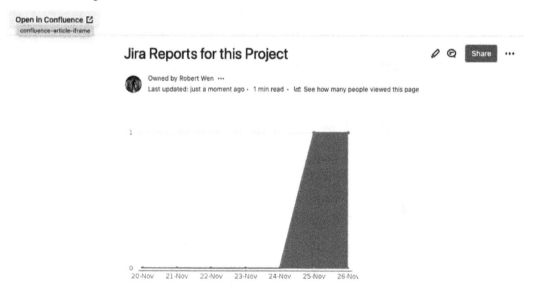

Insert JIRA Chart

Pie Chart	Search
Created vs Resolved	project="Open DevOps" Preview
Two Dimensional	Search using any issue key, search URL, JIRA link, JQL, plain text or filter, e.g. filter = "My JIRA filter"
OTHER JIRA CONTENT JIRA Issue/Filter	

1

⋙ **Display options**

Period   Daily   ⬍

Days previously*   7   ⑦

☐ Cumulative totals   ⑦
☐ Show unresolved trend   ⑦

Show versions   All versions   ⬍   ⑦

confluence-article-iframe   Width   [   ]   ⑦

☐ Show border
☐ Show chart information

Insert   Cancel

Figure 3.19 – Setting the configuration for the JIRA Charts macro

7. The chart will be generated and added to the Confluence page. The resulting page should look something like this:

Open in Confluence ⬏
confluence-article-iframe

Jira Reports for this Project

Owned by Robert Wen ⋯
Last updated: just a moment ago · 1 min read · ⬚ See how many people viewed this page

Figure 3.20 – A Jira chart on a Confluence page

By following the preceding steps, you have created and configured a Jira chart and set its display to a Confluence page.

Other reports use the **JIRA report** template.

Using the JIRA report template

Let's look at an example of creating a Confluence page with a template and configuring a report. Reports such as a change log or a status report can form the basis of design documentation that gets updated throughout the project. Follow these steps:

1. Create a new Confluence page using the **JIRA report** template. You can search for this by using the **More templates** option in the **Create a page** section.

Create a page

POPULAR TEMPLATES

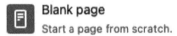

Blank page
Start a page from scratch.

Design decision
Record important project decisions and communicate them with your team.

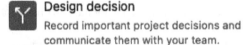

Meeting notes
Set meeting agendas, take notes, and share action items with your team.

Product requirements
Define, track and scope requirements for your product or feature.

Retrospective
What went well? What could have gone better? Crowdsource improvements with your team.

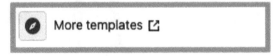

More templates ⌯

Figure 3.21 – Using the More templates option

2. On the page that opens, type Jira report in the search bar. You should see the following result:

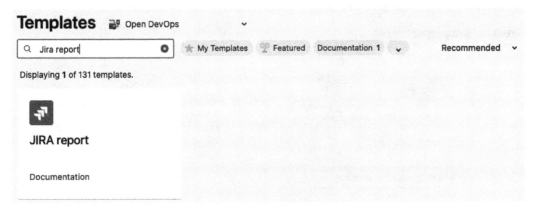

Figure 3.22 – Finding the JIRA report template

3. Selecting the **JIRA report** template and clicking **Use** opens a modal where you can select a report type, as shown here:

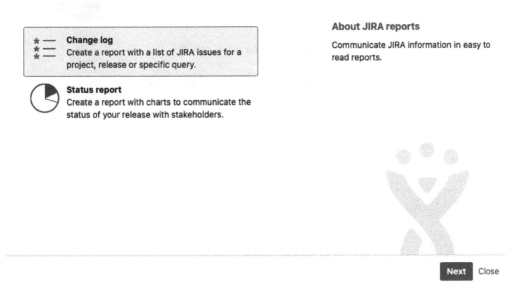

Figure 3.23 – Select report type

4.   Selecting the report type you wish to use and clicking **Next** opens another page where you can choose which project you want to create a status report for and edit its title.

Create a status report

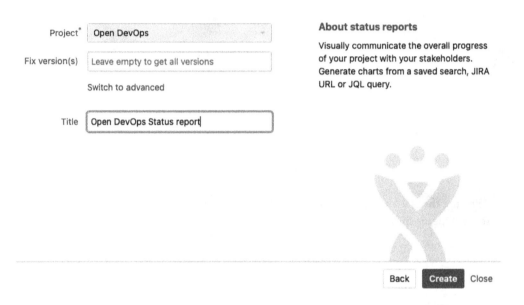

Figure 3.24 – Create a status report

5.   In the preceding figure, we selected a status report and identified its project. The respective page is created after clicking the **Create** button.

With that, we've created either a Jira change log or status report that's visible on a Confluence page based on the **JIRA report** template.

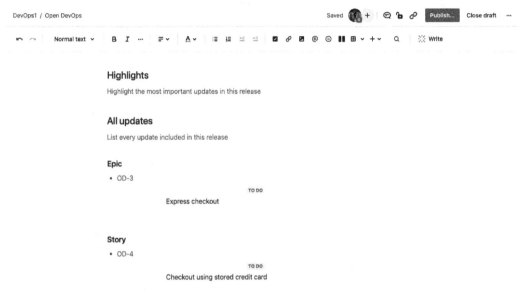

Figure 3.25 – Our newly created change log

Viewing Jira roadmaps on a Confluence page

If you've created a timeline in your project, you can display that timeline as a roadmap on a Confluence page. We will use the **Jira timeline** macro to facilitate this.

How to do it...

Let's learn how to create a Jira roadmap on a Confluence page:

1. On a Confluence page, look for the Jira timeline macro. Once found, add it.

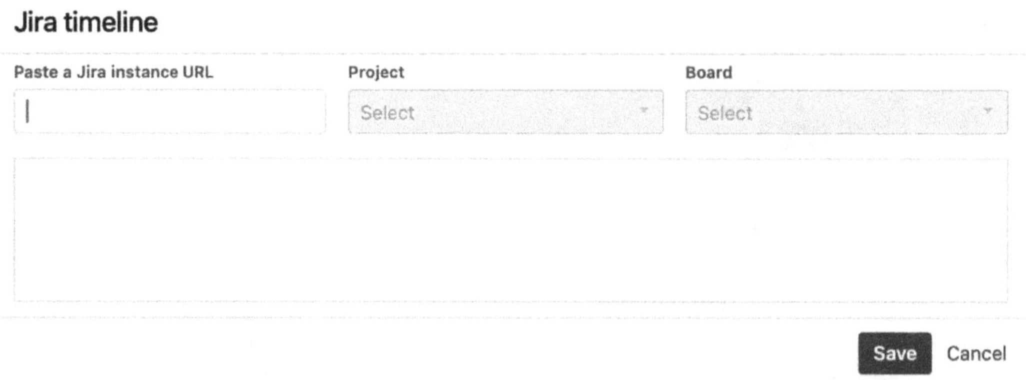

Jira timeline

Paste a Jira instance URL	Project	Board
	Select ▾	Select ▾

Save   Cancel

Figure 3.26 – Configuring the Jira timeline macro

2. Go to the **Jira timeline** page of your project, copy and paste the URL into the **Paste a Jira instance URL** text area, and click **Save**.

By pasting the URL of the **Jira timeline** page, you've linked the Jira roadmap to a Confluence page.

So far, we've been looking at connecting Jira to Confluence to allow Confluence to display Jira artifacts. In the next recipe, we'll turn our attention to connecting other applications so that they can do the same thing.

Viewing linked pages from other applications using Smart Links

Content that resides on other applications, such as Bitbucket, GitLab, and Figma, can be displayed dynamically on a Confluence page using **Smart Links**. Smart Links allow Confluence to display URL links from various applications, including Atlassian products such as Bitbucket and Jira, inline or embedded in pages.

Let's take a close look at how to embed linked page content onto a Confluence page using Smart Links.

Getting ready

Smart Links can be established from both Atlassian and non-Atlassian products. The following Atlassian products can have their links specified as Smart Links:

- Atlas
- Atlassian Analytics
- Bitbucket
- Compass
- Jira
- Jira Roadmaps
- Jira Align
- Trello

Non-Atlassian products that are commonly used in the application development process and have their links expressed as Smart Links in Confluence include Azure DevOps, GitHub, and GitLab. A complete list of supported applications can be found at `https://support.atlassian.com/confluence-cloud/docs/insert-links-and-anchors/#Smart-Links-from-Jira-and-other-products`.

Let's take a closer look at what's involved in setting up Smart Links.

Authenticating with the other application

For any non-Atlassian product that requires authentication, you must log into that application first to display the Smart Link on the Confluence page. You should only need to log in once. This will be effective on all sites in that Atlassian Cloud organization.

How to do it...

With authentication established, let's create and configure our Smart Links:

1. With the Confluence page in edit mode, copy the desired URL from the second application and paste it into the desired spot on your Confluence page.

2.   To make adjustments to the formatting of the linked content, click the Smart Link. Various controls should appear, as shown in the following screenshot:

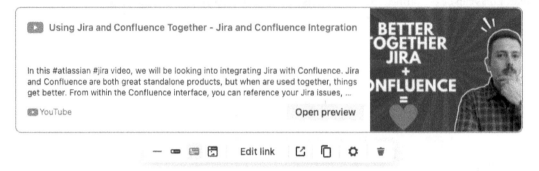

Figure 3.27 – Smart Link controls

3.   The first four icons determine how the link will be displayed. The order from left to right is as follows:

I.   Raw URL

II.   Title of the linked page

III. Card mode showing the title of the linked page and a description (shown in the preceding screenshot)

IV.   Embedded mode, where the contents of the linked page will be visible

The other controls allow you to perform the following actions:

I.   Edit the underlying URL link

II.   Open the linked page on another browser tab

III. Copy the URL link

IV.   Open the **Link preferences** page

V.   Remove the Smart Link from the Confluence page

Let's take a closer look at link preferences.

Working with link preferences

Link preferences are set on a per-account basis – that is, they are the preferred settings for all Smart Links that are created by an individual user. As such, they will show up as part of a user's Atlassian account settings, found under `https://id.atlassian.com`. Let's take a closer look:

1.   On the aforementioned page, you will find the following settings:

•   **Default display**: Sets the default display when creating a Smart Link

- **Exceptions for specific domains**: Specifies exceptions based on the link's domain

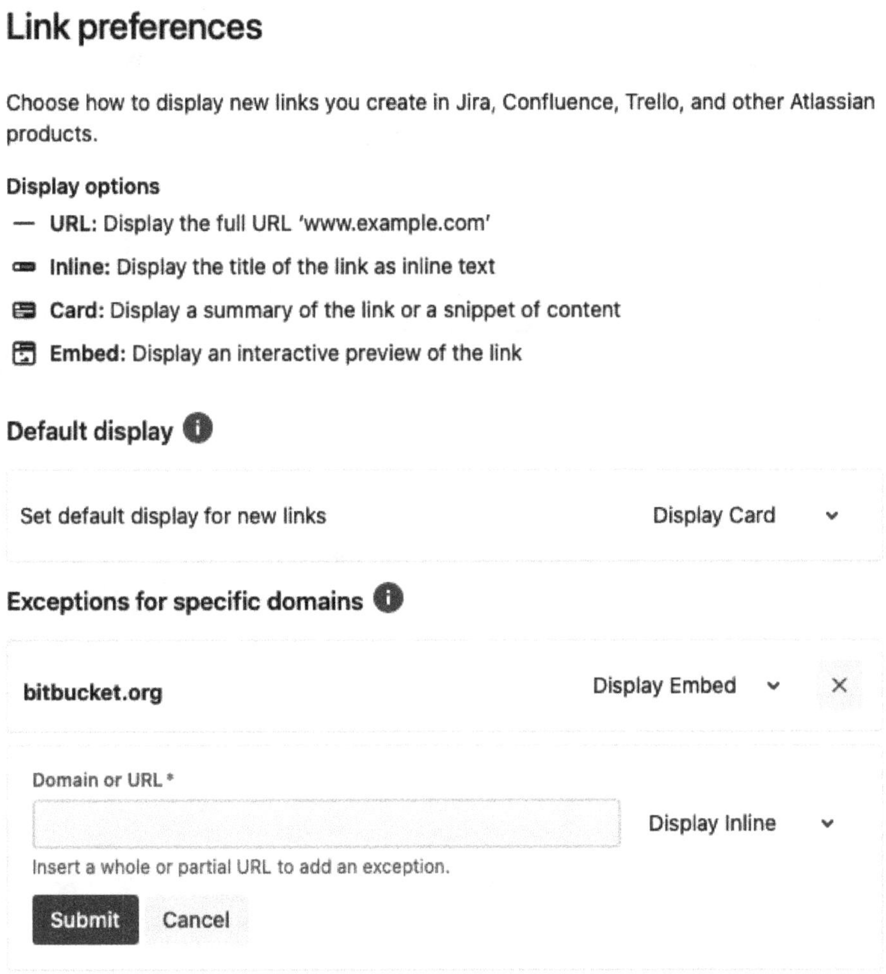

Link preferences

Choose how to display new links you create in Jira, Confluence, Trello, and other Atlassian products.

Display options

— **URL:** Display the full URL 'www.example.com'

▭ **Inline:** Display the title of the link as inline text

▤ **Card:** Display a summary of the link or a snippet of content

▥ **Embed:** Display an interactive preview of the link

Default display ⓘ

Set default display for new links	Display Card ⌄

Exceptions for specific domains ⓘ

bitbucket.org	Display Embed ⌄ ✕

Domain or URL *

	Display Inline ⌄

Insert a whole or partial URL to add an exception.

Submit   Cancel

Figure 3.28 – The Link preferences page

2. To set the default display, select the desired mode (**Display URL**, **Display Inline**, or **Display Card**).

3. To set an exception for the default display, type the domain's URL, select the desired mode (**Display URL**, **Display Inline**, **Display Card**, or **Display Embed**), and click **Submit**.

With that, we've set up the configuration that Atlassian will use to establish Smart Links when they're created by an individual, as well as how to create Smart Links.

Part 2: Development to Deployment

At this part of the development process, we are developing our implementation. That requires the software being developed to be under version control, where we track all changes against the previous versions.

Upon a commit performed in Git, a pipeline springs into action. Scanning and tests are run to find errors and security vulnerabilities. If everything passes, packaging of the build artifacts is done by the pipeline for deployment.

This part examines the role played by pipelines to perform all the necessary activities triggered by a commit to Git. We begin by describing how to connect Jira to external Git server tools and external pipeline tools for continuous integration.

We then introduce Bitbucket Cloud. This tool can act as both a Git server and a pipeline tool with Bitbucket Pipelines. With Bitbucket Pipelines, we explore the different functions available, including connecting with testing and security tools and, ultimately, deployment to different environments.

This part has the following chapters:

- *Chapter 4, Enabling Connections for Design, Source Control, and Continuous Integration*
- *Chapter 5, Understanding Bitbucket and Bitbucket Pipelines*
- *Chapter 6, Extending and Executing Bitbucket Pipelines*
- *Chapter 7, Leveraging Test Case Management and Security Tools for DevSecOps*
- *Chapter 8, Deploying with Bitbucket Pipelines*
- *Chapter 9, Leveraging Docker and Kubernetes for Advanced Configurations*

4

Enabling Connections for Design, Source Control, and Continuous Integration

The power of Atlassian's Open DevOps toolchain is the ability to integrate with many external third-party applications. While an organization may use Atlassian as the DevOps backbone, there may be other preferred applications for design, source control, or **continuous integration** (**CI**). For example, although Atlassian allows for easy connection to Bitbucket for **source control** and CI, we may want to connect to our existing tools, such as GitHub and GitLab for source control and Jenkins and CircleCI for CI. This chapter will demonstrate how to integrate Jira with these external tools to allow visibility of source control and CI information in Jira.

To accomplish this, we will look at incorporating the following recipes:

- Connecting Jira to design tools
- Connecting Jira to source control using a native integration
- Connecting Jira to source control using a universal integration
- Connecting Jira to CI tools

Technical requirements

To complete this chapter, you will need the following:

- Jira
- A Moqups account at https://moqups.com/
- A Figma account at https://www.figma.com/
- A GitLab account at https://gitlab.com/

- A CircleCI account at `https://circleci.com/`
- A local Jenkins server available from `https://www.jenkins.io/`

Connecting Jira to design tools

Visual design tools allow designers to create wireframes, mockups, and prototypes efficiently. By creating these detailed prototypes, you can identify usability issues and make necessary changes before any code is written. For example, if an organization wants to develop a mobile application, these design tools can be used to create a mockup of the UI. This mockup can then be attached to **Jira issues** that have been created to track this design work item.

Getting ready

To execute this recipe, you will need the following:

- Jira administration permissions in order to install the necessary app plugins
- Your Moqups account
- Your Figma account

How to do it...

This recipe has the following objectives:

- Configuring the integration of two popular design tools (Moqups and Figma) with Jira
- Demonstrating the usage of design tools in Jira issues

Let's proceed to achieving these objectives.

How to connect Jira to Moqups

Moqups is a web-based application designed to facilitate the creation of wireframes, mockups, diagrams, and prototypes. These are essential components of the early stages of website, software, and application development processes.

Let's get started on adding the Moqups application to Jira.

1. Go to the application marketplace and search for the Moqups app. Select the **Moqups for Jira Cloud** application.

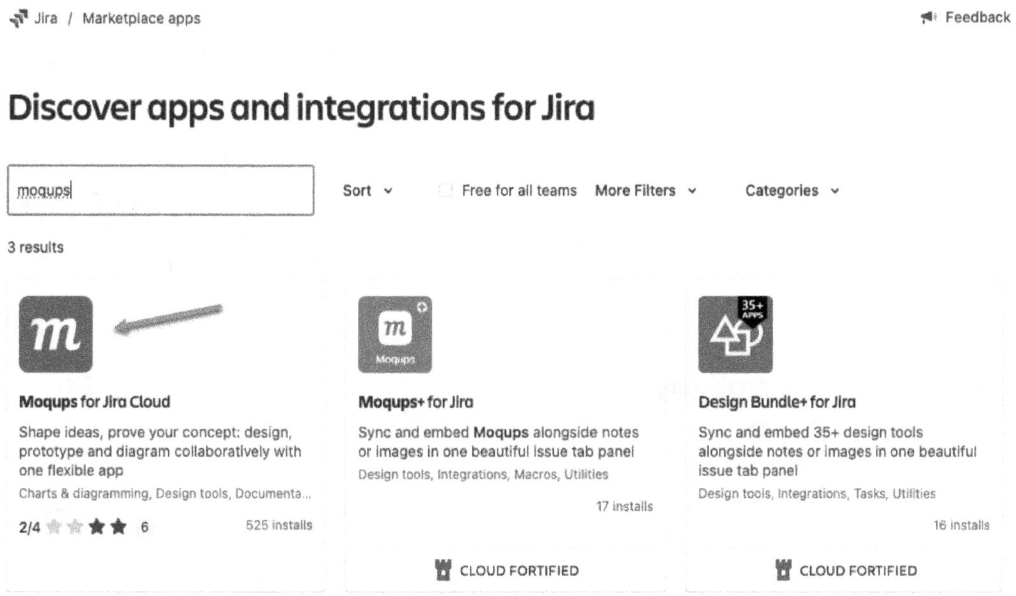

Figure 4.1 – Finding the Moqups application

2. An application information page is displayed. Select **Get app** to begin the installation process.

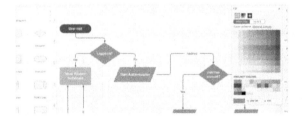

Figure 4.2 – Getting the Moqups application

3. An **Add to Jira** pop-up window will present itself. Select the **Get it now** button to continue the installation process.

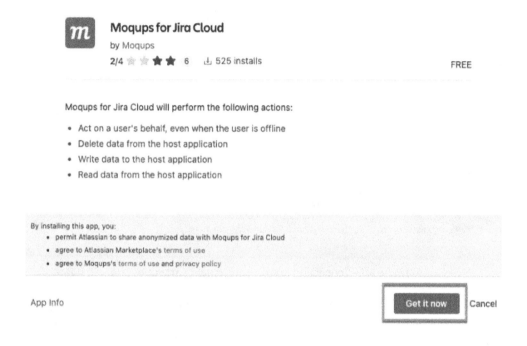

Figure 4.3 – Adding application pop-up window

4. Once the application is installed and ready, a window will display, allowing you to configure the application. Select the **Configure** option.

Figure 4.4 – Configuring Moqups application

5. To configure the application, you must connect to your existing Moqups account. This will establish an integration, allowing you to pull in any of your existing Moqups projects. Select the **Connect to an existing Moqups account** button.

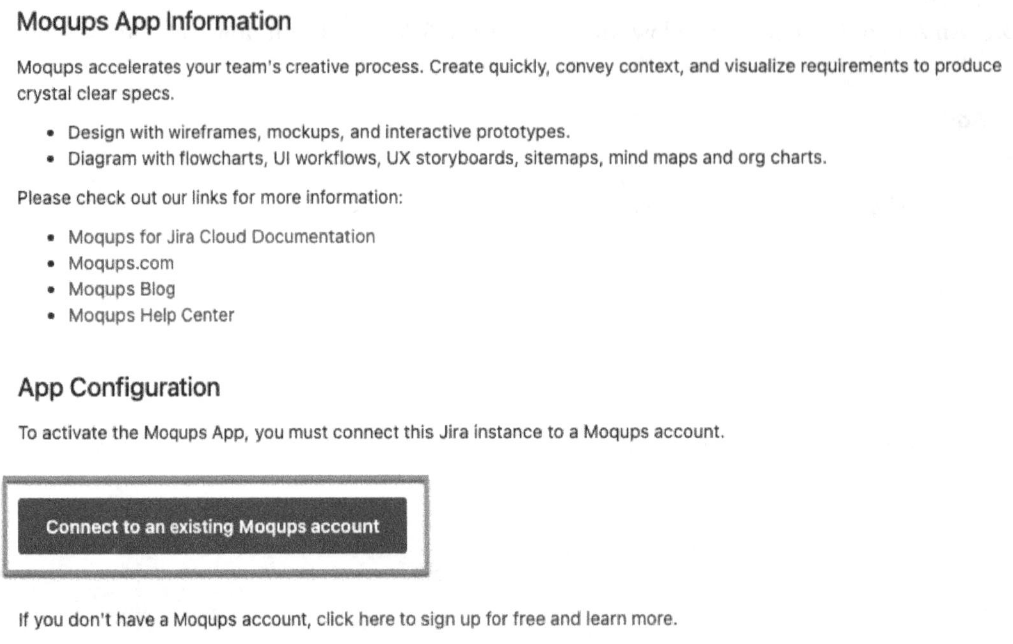

Moqups App Information

Moqups accelerates your team's creative process. Create quickly, convey context, and visualize requirements to produce crystal clear specs.

- Design with wireframes, mockups, and interactive prototypes.
- Diagram with flowcharts, UI workflows, UX storyboards, sitemaps, mind maps and org charts.

Please check out our links for more information:

- Moqups for Jira Cloud Documentation
- Moqups.com
- Moqups Blog
- Moqups Help Center

App Configuration

To activate the Moqups App, you must connect this Jira instance to a Moqups account.

> **Connect to an existing Moqups account**

If you don't have a Moqups account, click here to sign up for free and learn more.

Figure 4.5 – Connecting to an existing Moqups account

6. A Moqups credential popup will be displayed. Enter your Moqups account credentials and select the **Connect** button.

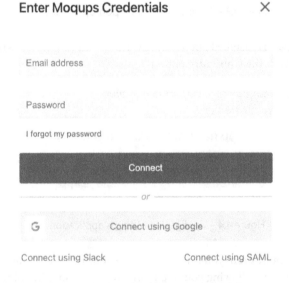

Enter Moqups Credentials   ✕

Email address

Password

I forgot my password

Connect

————————————— or —————————————

G   Connect using Google

Connect using Slack   Connect using SAML

Figure 4.6 – Entering Moqups credentials

7. Once your Moqups credentials have been successfully validated, you will be presented with the **Moqups App Information** page. Here, you can adjust permissions, subscription quota, and/or disconnect from Moqups.

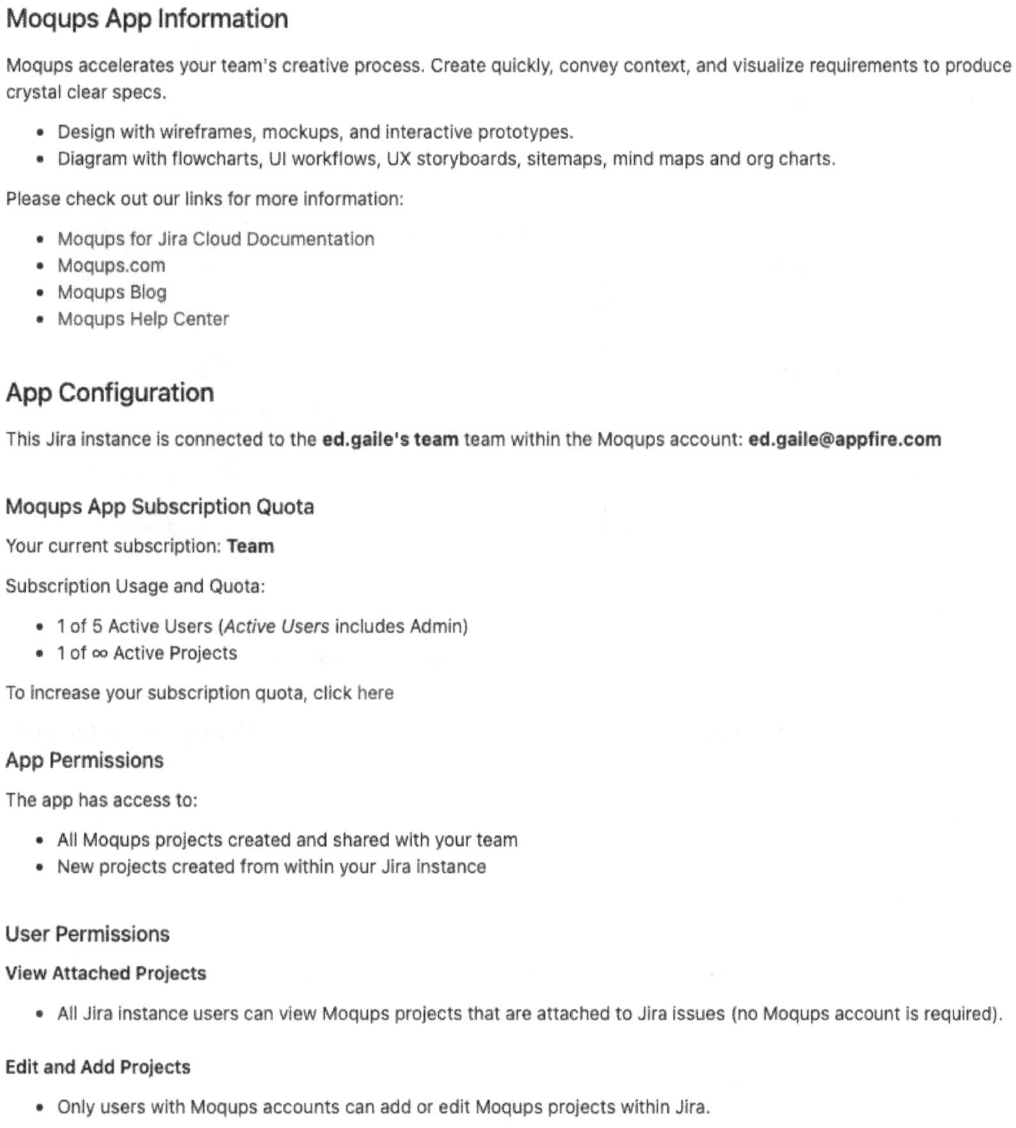

Moqups App Information

Moqups accelerates your team's creative process. Create quickly, convey context, and visualize requirements to produce crystal clear specs.

- Design with wireframes, mockups, and interactive prototypes.
- Diagram with flowcharts, UI workflows, UX storyboards, sitemaps, mind maps and org charts.

Please check out our links for more information:

- Moqups for Jira Cloud Documentation
- Moqups.com
- Moqups Blog
- Moqups Help Center

App Configuration

This Jira instance is connected to the **ed.gaile's team** team within the Moqups account: **ed.gaile@appfire.com**

Moqups App Subscription Quota

Your current subscription: **Team**

Subscription Usage and Quota:

- 1 of 5 Active Users (*Active Users* includes Admin)
- 1 of ∞ Active Projects

To increase your subscription quota, click here

App Permissions

The app has access to:

- All Moqups projects created and shared with your team
- New projects created from within your Jira instance

User Permissions

View Attached Projects

- All Jira instance users can view Moqups projects that are attached to Jira issues (no Moqups account is required).

Edit and Add Projects

- Only users with Moqups accounts can add or edit Moqups projects within Jira.

Figure 4.7 – Moqups configuration settings

8. Now that Moqups is installed and configured, we can add a project to any Jira issue. To accomplish this, go to an appropriate Jira issue and select the three dots for additional options. Select the **Moqups** option.

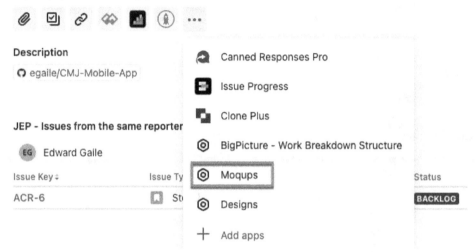

Figure 4.8 – Selecting Moqups to add a component to Jira

9. A **Moqups** section will be provided on the Jira issue. Select the **Add Moqups Project** to pull a project into the issue.

Figure 4.9 – Adding a Moqups project to the Jira issue

10. All available Moqups projects will be available to the user based on their permissions. Select the appropriate project and then click the **Add** button.

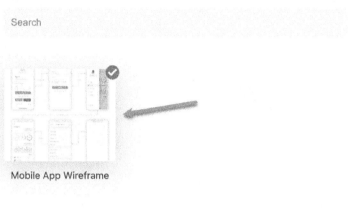
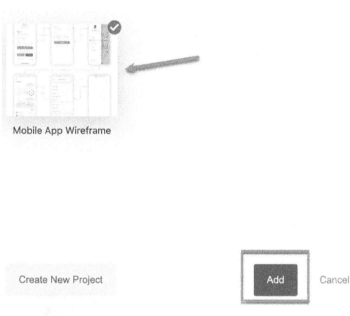

Figure 4.10 – Selecting and adding the Moqups project

11. The Moqups project is now embedded as an object in the **Moqups** section of the Jira issue.

Figure 4.11 – Moqups project on Jira ticket

12. You can now edit the Moqups project right from the Jira ticket. Select the pencil icon from the **Moqups** object, and you will be taken to that project in edit mode.

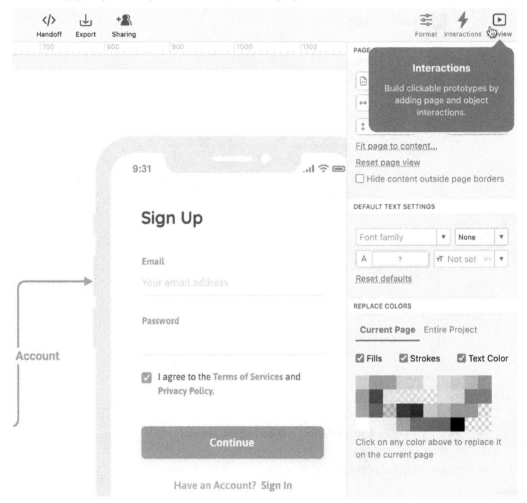

Figure 4.12 – Editing the Moqups project

Great job! You have now achieved the objectives of this recipe by integrating the Moqups design app with Jira. You can now keep your design and project management teams fully in synch.

Next up, we are going to look at doing a similar activity but using the Figma design application.

How to connect Jira to Figma

Figma is a cloud-based design tool that is widely used for its collaborative and user-friendly interface. It is primarily aimed at UI and UX design for applications, but it also has robust capabilities for graphic design and prototyping.

Let's now add the Figma app to Jira to demonstrate another design-type application.

1. Go to the application marketplace and search for the Figma app. Select the **Figma for Jira** app that has the Figma logo.

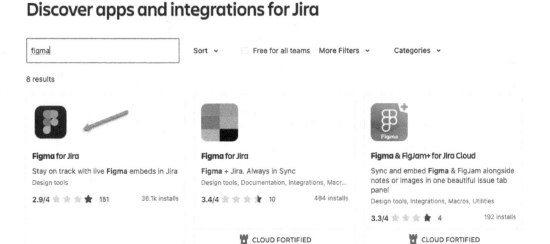

Figure 4.13 – Installing the Figma app

2. An application information page is displayed. Select **Get app** to begin the installation process.

Figure 4.14 – Installing the Figma app

3. An **Add to Jira** pop-up window will present itself. Select the **Get it now** button to continue the installation process.

Add to Jira

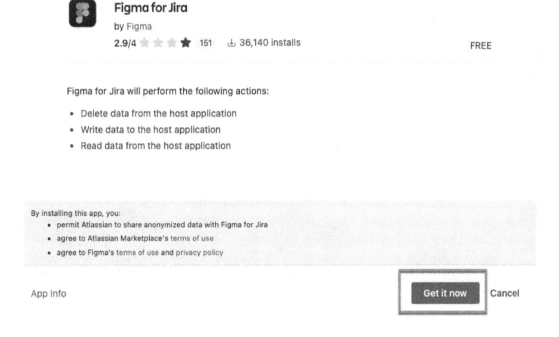

Figure 4.15 – Installing the Figma app

4. Now, switch over to your existing Figma account and select the project you would like to share.

5. Update the project permissions to allow users to view and interact with the file. Select **Copy link** to get the file URL.

Figure 4.16 – Sharing the Figma project

6. Switch back to Jira and select the Jira issue you want to add to the file embed. Select the three dots for more options and select **Designs**.

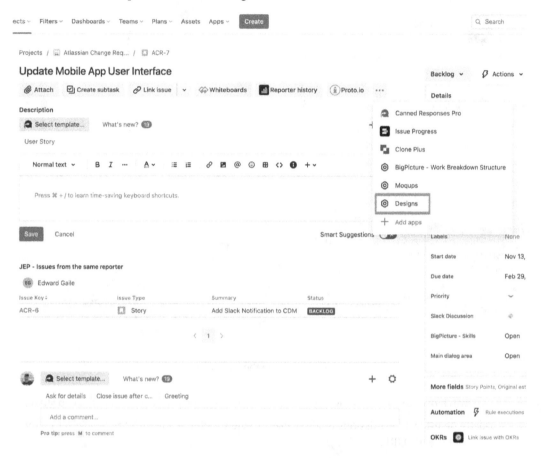

Figure 4.17 – Selecting Designs from the menu

7. A **Designs** section will appear on the Jira issue. Paste the copied Figma file URL and select **Add Design**.

Figure 4.18 – Adding Figma public URL

8. The Figma project is now added to the Jira issue in the **Designs** section:

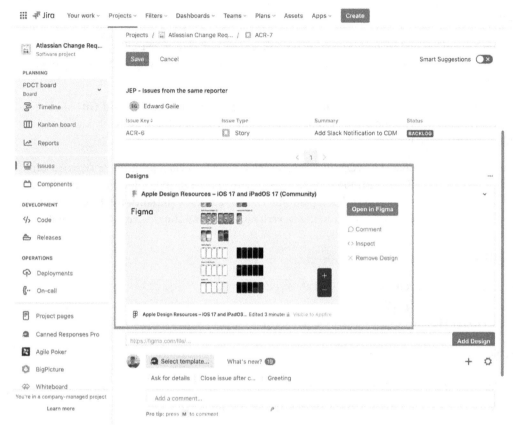

Figure 4.19 – Figma project now embedded in Jira issue

You have successfully added the Figma integration to Jira! This recipe has shown how we can incorporate design activities into our Jira issue management.

Next up, we look at adding source control applications via native integration.

Connecting Jira to source control using a native integration

Most source control platforms have two ways to connect to Jira: a native connection (from the perspective of the source control application) or a plugin installation using Atlassian's **Universal Plugin Manager** (**UPM**).

In this section, we will look at how to use **native integration** to connect Jira and **GitLab**. Native integration means we are adding the integration from the GitLab application directly and not going through an app on the Jira side.

GitLab is a popular web-based DevOps life-cycle tool that provides a Git repository for source control. GitLab also features wiki, CI/**continuous deployment/delivery** (**CD**), and code review capabilities. It is a complete application that allows developers and teams to cover the full **software development life cycle** (**SDLC**), from planning to creation, building, verifying, security testing, deploying, and monitoring.

Getting ready

To execute this recipe, you will need a GitLab account at `https://gitlab.com/`.

How to do it...

The objectives for this recipe are the following:

- Configuring a native GitLab integration with Jira
- Verifying the connection by viewing associated Jira issues in GitLab

To achieve these objectives, we will use the following steps:

1. Connect Jira with GitLab through the native integration, go to your GitLab account, and select the desired project:

Figure 4.20 – GitLab project

2. From the right-hand menu options, select **Settings**.

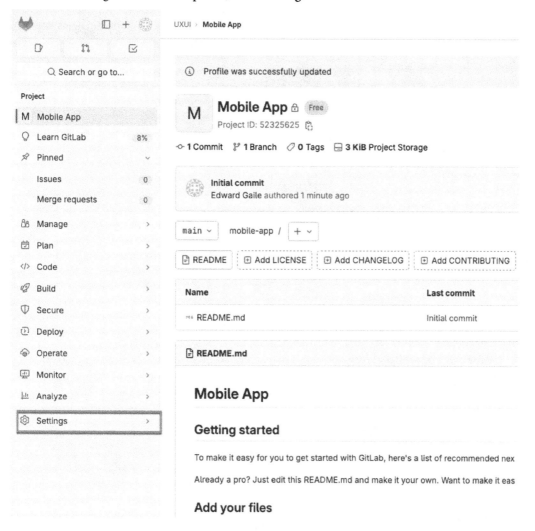

Figure 4.21 – GitLab project settings

3. This will present a sub-menu. From here, select the **Integrations** option.

Figure 4.22 – GitLab project integrations

4. A full page of available integrations is displayed. Scroll down to the Jira integration and select **Configure**.

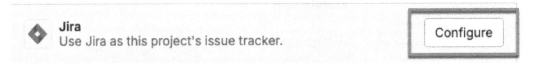

Figure 4.23 – Jira integration configuration

5. Jira connection details will be displayed. To configure the Jira endpoint, complete the following actions.

    I.   Check the **Enable integration** option.

    II.   Under **Authentication method**, check **Basic**, then do the following:

          i.   Add your **Email or username** details.

ii.   Add your **API Token or password** details.

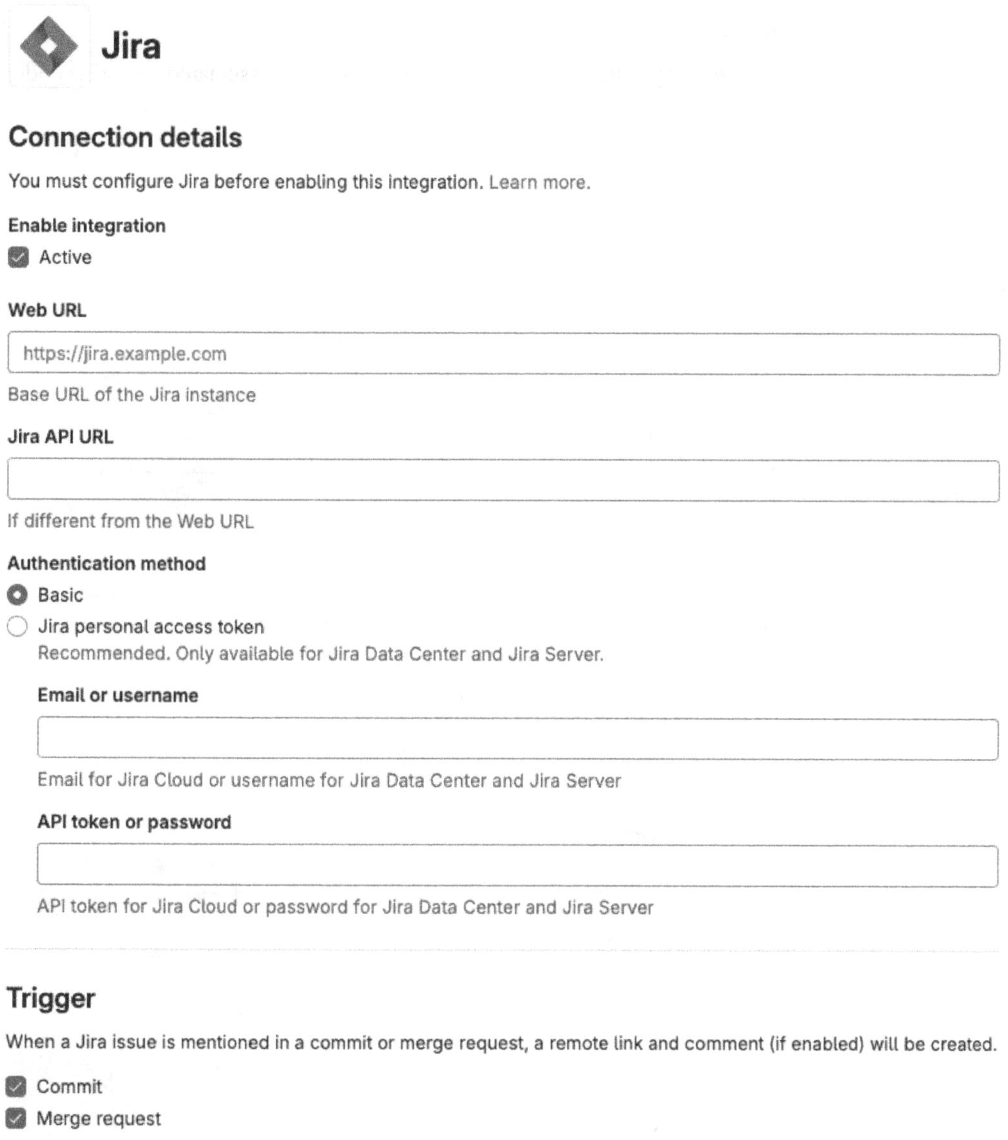

Jira

Connection details

You must configure Jira before enabling this integration. Learn more.

Enable integration

☑ Active

Web URL

```
https://jira.example.com
```

Base URL of the Jira instance

Jira API URL

If different from the Web URL

Authentication method

🔘 Basic

⭕ Jira personal access token
Recommended. Only available for Jira Data Center and Jira Server.

Email or username

Email for Jira Cloud or username for Jira Data Center and Jira Server

API token or password

API token for Jira Cloud or password for Jira Data Center and Jira Server

Trigger

When a Jira issue is mentioned in a commit or merge request, a remote link and comment (if enabled) will be created.

☑ Commit
☑ Merge request

Figure 4.24 – Jira connection parameters

6. Scroll down to the bottom of the Jira **Connection details** page and perform the following actions:

I. Check **Enable Jira issues**.

II. Add the **Jira project key** value for the project you want to be associated with this code repository project.

III. Select **Save changes**.

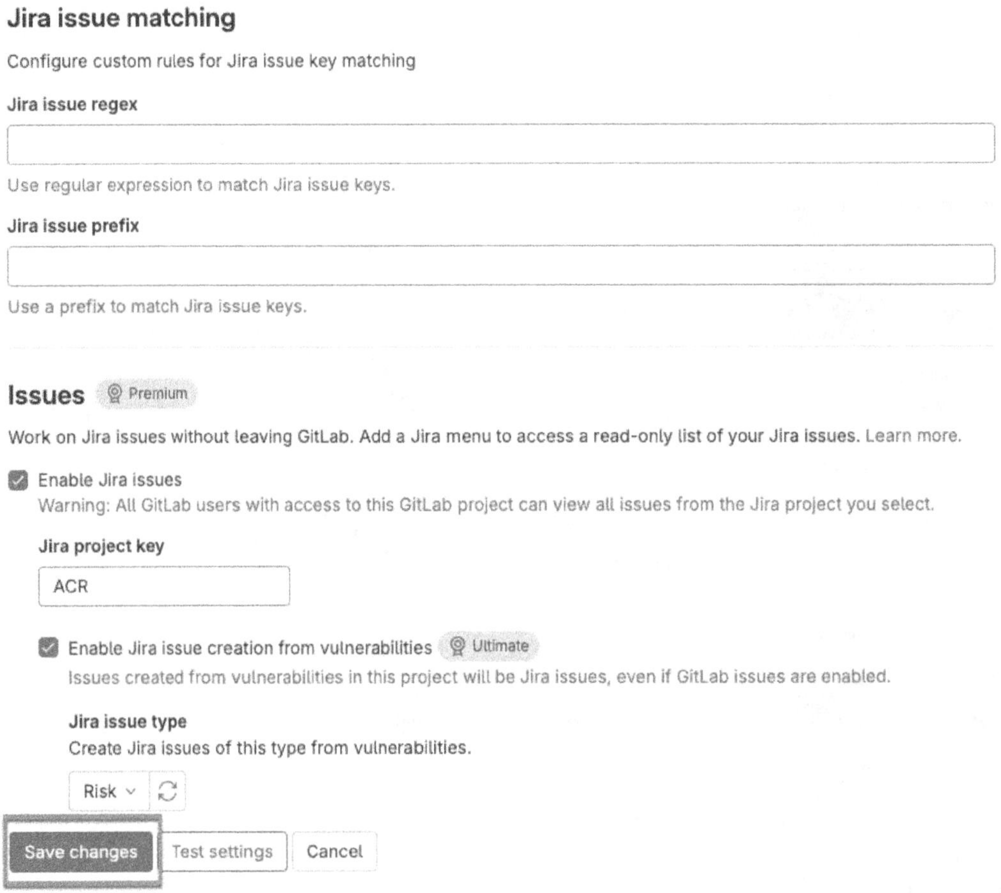

Jira issue matching

Configure custom rules for Jira issue key matching

Jira issue regex

Use regular expression to match Jira issue keys.

Jira issue prefix

Use a prefix to match Jira issue keys.

Issues ⓟ Premium

Work on Jira issues without leaving GitLab. Add a Jira menu to access a read-only list of your Jira issues. Learn more.

☑ Enable Jira issues
    Warning: All GitLab users with access to this GitLab project can view all issues from the Jira project you select.

    Jira project key

    ACR

    ☑ Enable Jira issue creation from vulnerabilities ⓤ Ultimate
        Issues created from vulnerabilities in this project will be Jira issues, even if GitLab issues are enabled.

        Jira issue type
        Create Jira issues of this type from vulnerabilities.

        Risk ∨ ⟳

Save changes   Test settings   Cancel

Figure 4.25 – Jira issue enablement

7. On the left-hand menu options, you can now select **Plan**, which will produce a sub-menu with the **Jira issues** option. Select **Jira issues**.

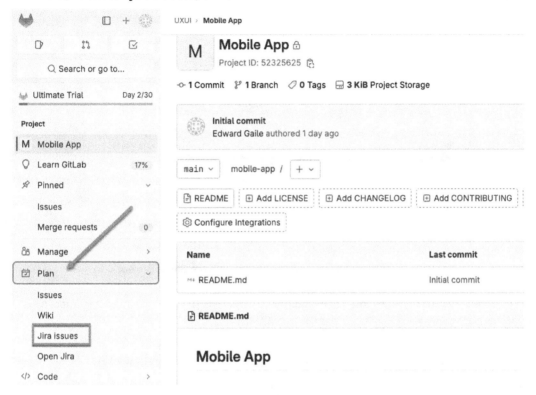

Figure 4.26 – Selecting associated Jira issues

8. You will now see the Jira issues the project key associated with this GitLab repository.

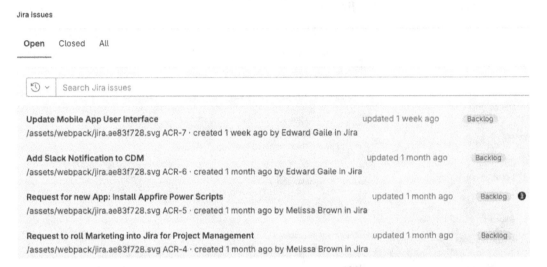

Figure 4.27 – Viewing Jira issues associated with the GitLab project

You have now successfully added the Jira integration to GitLab via the GitLab native integration. Next up, we look at adding source control from the Jira UPM.

Connecting Jira to source control using a universal integration

One of the easiest ways to connect source control to Jira is through Atlassian's UPM.

In this section, you will learn how to establish connections to an external source control tool such as **GitHub** and view connected Git repo data from Jira using the GitHub for Jira app.

GitHub is a cloud-based platform that provides source control for software development using Git. GitHub also provides developers with bug tracking, task management, and CI/CD capabilities.

Getting ready

To execute this recipe, you will need the following:

*   Jira administration permissions to install the GitHub for Jira app
*   A GitHub account at https://github.com/

How to do it...

The objectives for this recipe are to install and configure the GitHub for Jira application. To do that, use the following steps:

1. First, we need to install the free GitHub app into Jira.

    Go to the application marketplace and search for the GitHub app. Select the **GitHub for Jira** app from Atlassian.

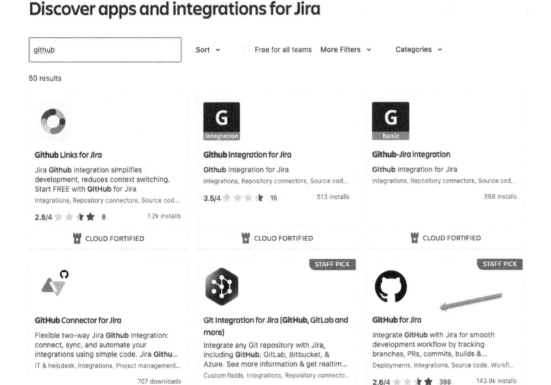

Figure 4.28 – Finding the GitHub for Jira app

2.   An application information page is displayed. Select **Get app** to begin the installation process.

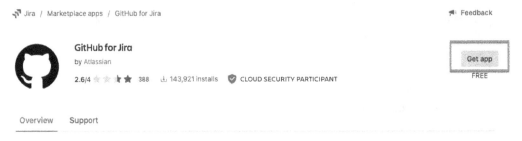

Figure 4.29 – Installing the GitHub app

3. An **Add to Jira** pop-up window will present itself. Select the **Get it now** button to continue the installation process.

Add to Jira

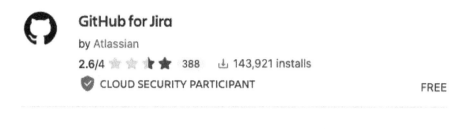

GitHub for Jira will perform the following actions:

- Delete data from the host application
- Write data to the host application
- Read data from the host application

App Info

Figure 4.30 – Adding GitHub app to Jira

4. When the GitHub app has finished being installed into Jira, a pop-up window will display, allowing you to configure the app by selecting **Get started**.

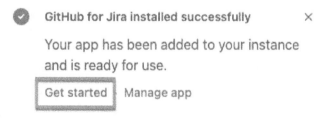

Figure 4.31 – Configuring the GitHub app

5.   A **Connect Github to Jira** screen is displayed. Select **Continue**.

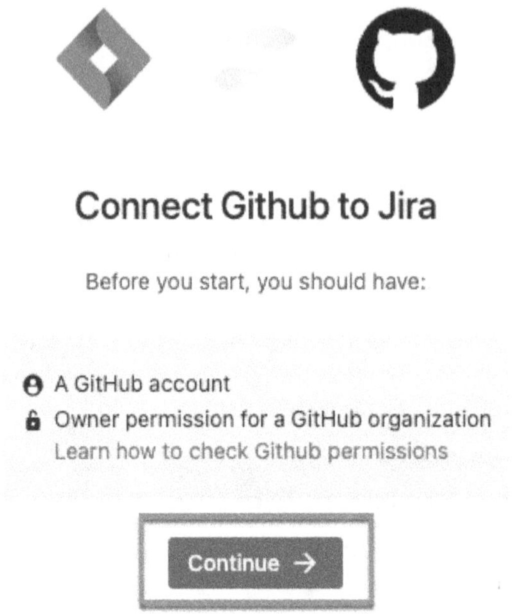

Connect Github to Jira

Before you start, you should have:

A GitHub account
Owner permission for a GitHub organization
Learn how to check Github permissions

Continue →

Figure 4.32 – GitHub connection

6.   Authorize the GitHub links by highlighting the **GitHub Cloud** option and selecting **Next**.

Connect Github to Jira

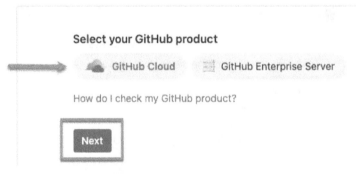

Select your GitHub product

GitHub Cloud   GitHub Enterprise Server

How do I check my GitHub product?

Next

Figure 4.33 – Selecting GitHub Cloud or GitHub Enterprise Server

7. Jira needs access authorization from GitHub. Select **Authorize Jira**.

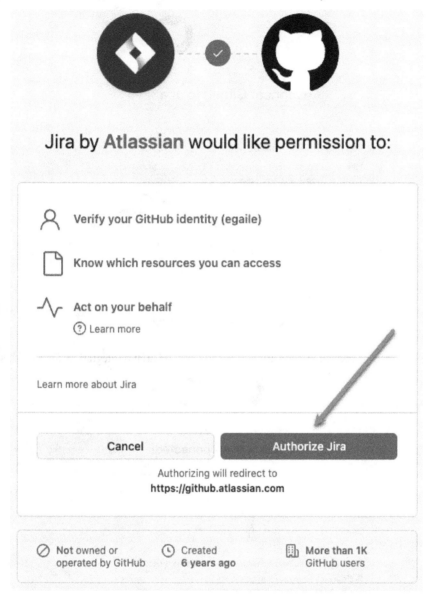

Figure 4.34 – GitHub configuration instructions

8.   The final step for GitHub access is to select **Connect**.

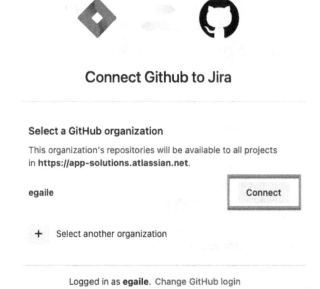

Figure 4.35 – Connecting GitHub

9.   A response page is presented, indicating successful Jira to GitHub integration. Select **Exit set up**.

Figure 4.36 – GitHub connection

10. A **GitHub configuration** page is returned showing the connected organizations available.

GitHub configuration

Connecting GitHub to Jira allows you to view development activity in the context of your Jira project and issues. To send development data from GitHub to Jira, your team must include iss request titles.

Even if your organization is still backfilling historical data, you can start using issue keys in your development work immediately.

Connected organization	Repository access	Backfill status ℹ
egaile	All repos 1 ✎	FINISHED Backfilled from: 5/25/23 ℹ

Figure 4.37 – Adding URL to Jira description

Congratulations – you have now integrated the GitHub source control through the UPM process!

Next up, we look at adding CI tools to Jira!

Connecting Jira to CI tools

In this section, we will explore integrating Jira with CI tools. Combining Jira and CI gives users a detailed view of development activities, including code commits, pull requests, and releases.

Getting ready

We need the following pre-requisites for both parts of this recipe:

- Jenkins server
- Atlassian Jira plugin installed on the Jenkins server
- CircleCI account
- Organization ID from the CircleCI account

How to do it...

We will look at two recipes for this section, starting with connecting Jira to Jenkins, followed by connecting Jira to CircleCI.

First up is Jenkins. Jenkins is a popular open source automation server that enables developers to build, test, and deploy their software projects continuously. A Jenkins server can easily be integrated with Jira.

How to connect Jira to Jenkins

Let's begin by adding the Jenkins app.

1.   Go to the application marketplace and search for the Jenkins app. Select the **Jenkins for Jira (Official)** app.

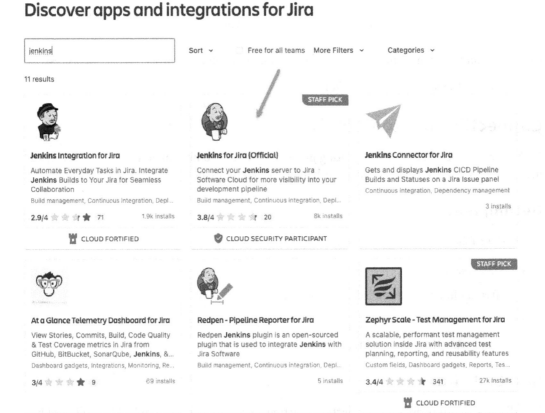

Figure 4.38 – Finding the Jenkins for Jira app

2.   An application information page is displayed. Select **Get app** to begin the installation process.

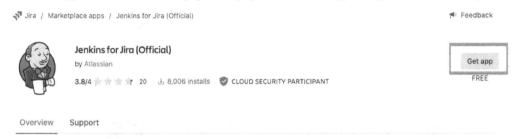

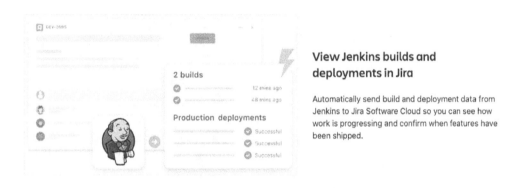

Connect your Jenkins server to Jira Software Cloud for more visibility into your development pipeline

Figure 4.39 – Getting Jenkins for Jira app

3.   An **Add to Jira** pop-up window will present itself. Select the **Get it now** button to continue the installation process.

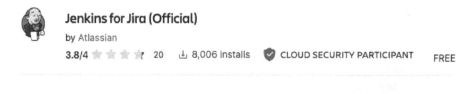

Add to Jira

Jenkins for Jira (Official)
by Atlassian

3.8/4 ⭐⭐⭐⭐⭐ 20   ⬇ 8,006 installs   🛡 CLOUD SECURITY PARTICIPANT   FREE

Jenkins for Jira (Official) will perform the following actions:

- Allows the app to delete build information
- Allows the app to delete deployment information
- View permissions.
- Read and write to app storage service

Expand all details

Jenkins for Jira (Official) can send data to the following domains:

- https://api-private.atlassian.com
- https://app.launchdarkly.com
- https://events.launchdarkly.com
- https://clientstream.launchdarkly.com

App Info   Get it now   Cancel

Figure 4.40 – Installing the Jenkins for Jira app

4. Once the Jenkins app is installed, we can proceed to configuration. Navigate to the Jira **Manage apps** page and select the **Jenkins for Jira** item from the left-hand menu.

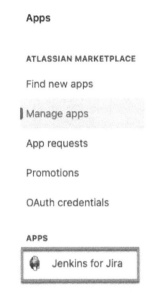

Figure 4.41 – Configuring the Jenkins for Jira app

5. A **Connect Jenkins to Jira Software** page is displayed. Select the **Connect a Jenkins server** button to continue configuration.

Figure 4.42 – Connecting to a Jenkins server

6.   The configuration will provide a notification that the **Atlassian Jira Software Cloud** plugin should already be installed on the Jenkins server. Go ahead and click the **Next** button.

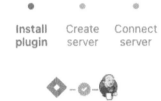

Before you continue, install plugin on Jenkins

On your Jenkins server, install the "Atlassian Jira Software Cloud" plugin. You must be an admin to do this.

1. Open your Jenkins server
2. Navigate to **Manage Jenkins > Manage plugins**
3. In the **Available** tab, search for "**Atlassian Jira Software Cloud**"
4. Check the "**Install**" checkbox
5. Click "**Download now and install after restart**"

Once you've installed the plugin on Jenkins, click "Next".

Figure 4.43 – Confirming Jira plugin installed on Jenkins

7. We can start to create a connection by first providing a name for our Jenkins server connection and then selecting **Create**.

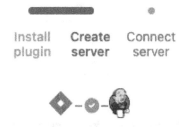

Install   **Create**   Connect
plugin   **server**   server

Create your Jenkins Server

Jenkins server name ❓

Server name

Local-Jenkins

Enter a name for your server. You can change this at any time.

Create

Figure 4.44 – Creating a Jenkins server connection

8. The Jenkins for Jira app will provide you with a Webhook URL and a secret key. These parameters are what you need to cut and paste into the **Connect Jenkins to your Jira site** configuration.

Select **Done**.

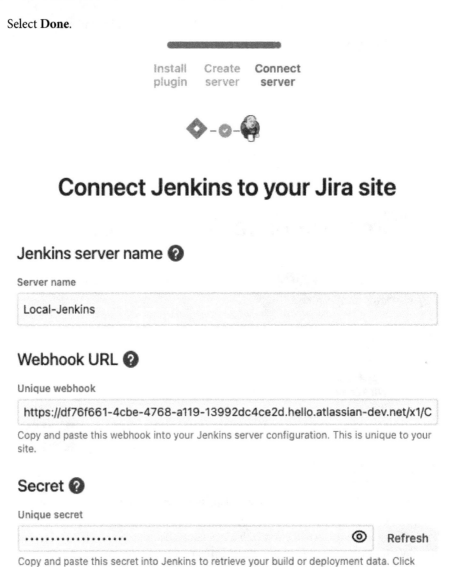

Figure 4.45 – Jenkins connection parameters

The Jenkins for Jira app now displays the connected Jenkins server.

Jenkins configuration

After you connect your Jenkins server to Jira and send a deployment event from your CI/CD tool, you will be able to view development info with insights.

Connected Jenkins servers

Last 0 active pipelines

Waiting for build or deployment event

Figure 4.46 – Jenkins connected

Jenkins is one of the most popular CI tools, and you have now successfully connected the application to your Jira instance.

Next up, we look at integrating another popular CI tool: CircleCI.

How to connect Jira to CircleCI

CircleCI is a modern CI/CD platform that automates the software development process, allowing teams to build, test, and deploy applications rapidly and reliably. It is designed to work with cloud-based or on-premise **version control systems** (**VCSs**) such as GitHub and Bitbucket.

Let's start adding the CircleCI application to Jira.

1. First, go to the application marketplace and search for the CircleCI app. Select the **CircleCI for Jira** app.

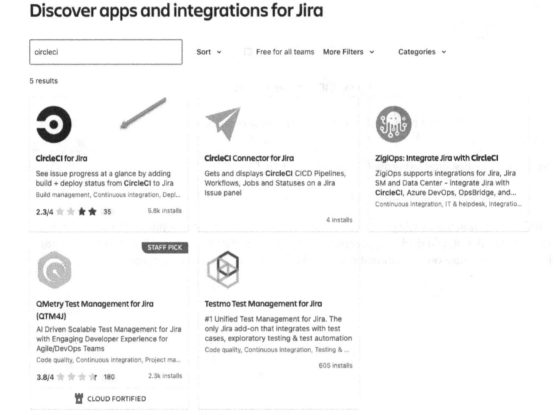

Figure 4.47 – Finding the CircleCI for Jira app

2. An application information page is displayed. Select **Get app** to begin the installation process.

Display build and deployment status in Jira issues

Connect CircleCI + Jira so everyone on your team can see the build, test, and deploy status of your issues, at a glance.

Figure 4.48 – Getting the CircleCI for Jira app

3. An **Add to Jira** pop-up window will present itself. Select the **Get it now** button to continue the installation process.

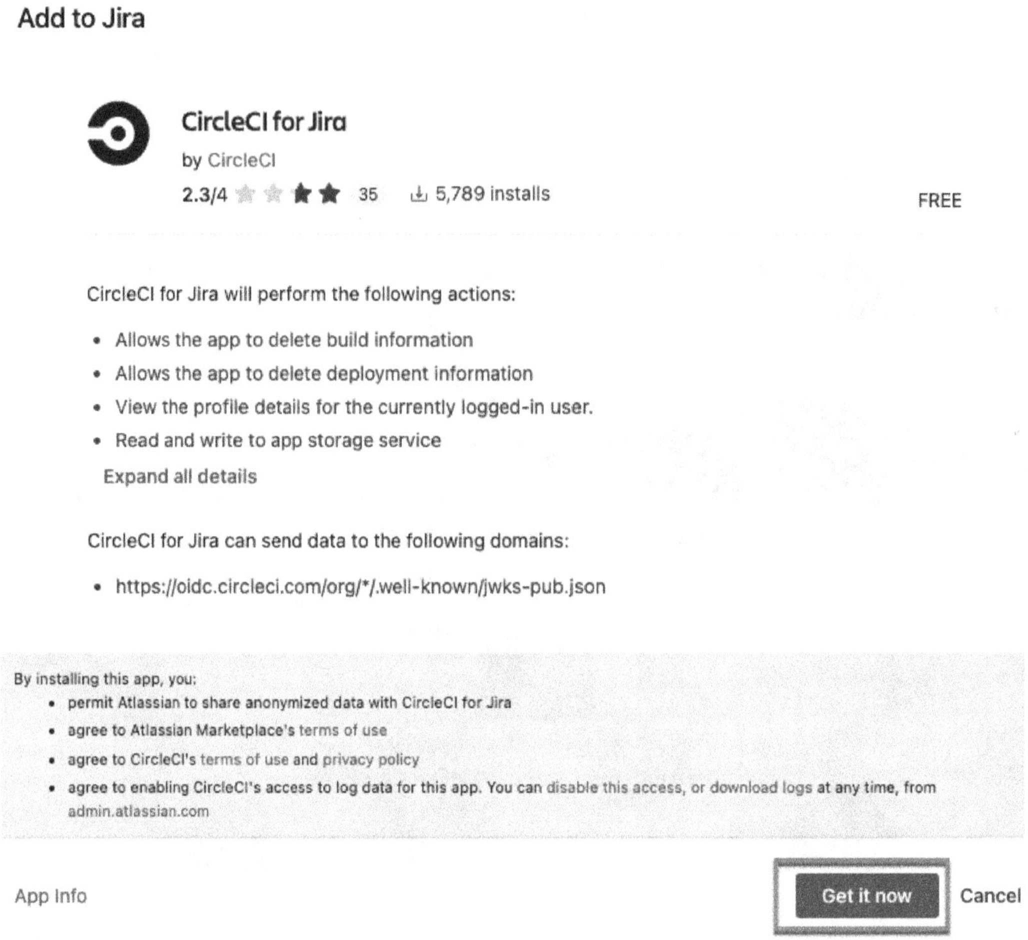

Add to Jira
CircleCI for Jira
by CircleCI
2.3/4
35
5,789 installs
FREE
CircleCI for Jira will perform the following actions:
Allows the app to delete build information
Allows the app to delete deployment information
View the profile details for the currently logged-in user.
Read and write to app storage service
Expand all details
CircleCI for Jira can send data to the following domains:
https://oidc.circleci.com/org/*/.well-known/jwks-pub.json
By installing this app, you:
permit Atlassian to share anonymized data with CircleCI for Jira
agree to Atlassian Marketplace's terms of use
agree to CircleCI's terms of use and privacy policy
agree to enabling CircleCI's access to log data for this app. You can disable this access, or download logs at any time, from admin.atlassian.com
App Info
Get it now
Cancel

Figure 4.49 – Installing the CircleCI for Jira app

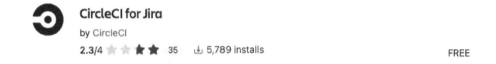

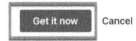

4. Once the CircleCI app is installed, we can proceed to configuration. Go to the Jira **Manage apps** page and select the **CircleCI for Jira** item from the left-hand menu.

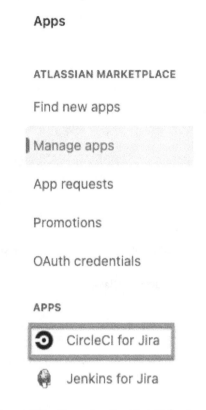

Figure 4.50 – Managing the CircleCI for Jira app

5.   A **CircleCI for Jira** configuration page is displayed; select **CONFIGURE**.

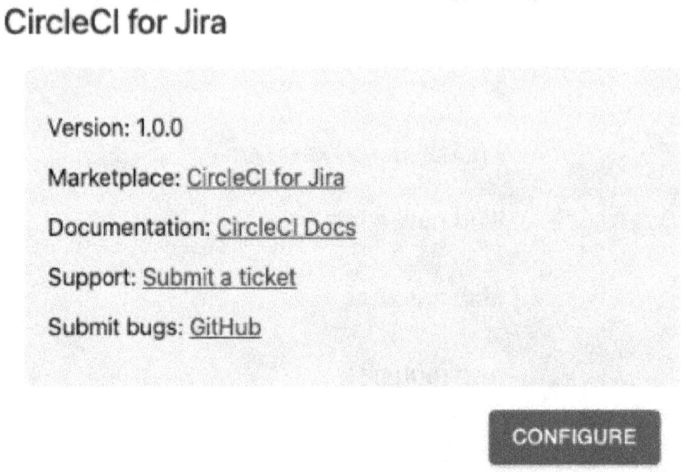

CircleCI for Jira

Version: 1.0.0

Marketplace: CircleCI for Jira

Documentation: CircleCI Docs

Support: Submit a ticket

Submit bugs: GitHub

CONFIGURE

Figure 4.51 – Configuring the CircleCI for Jira app

6.   CircleCI will ask for permission to access Atlassian products; select **Allow access**.

CircleCI for Jira - Configure

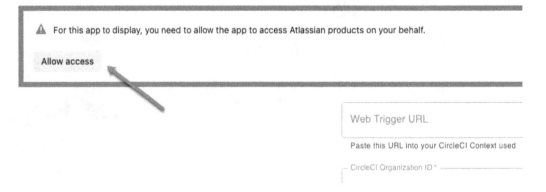

⚠ For this app to display, you need to allow the app to access Atlassian products on your behalf.

Allow access

Web Trigger URL

Paste this URL into your CircleCI Context used

CircleCI Organization ID *

Figure 4.52 – Allowing CircleCI access

7. Once Atlassian product access is granted, the CircleCI app will request access to your Atlassian account. Select **Accept**.

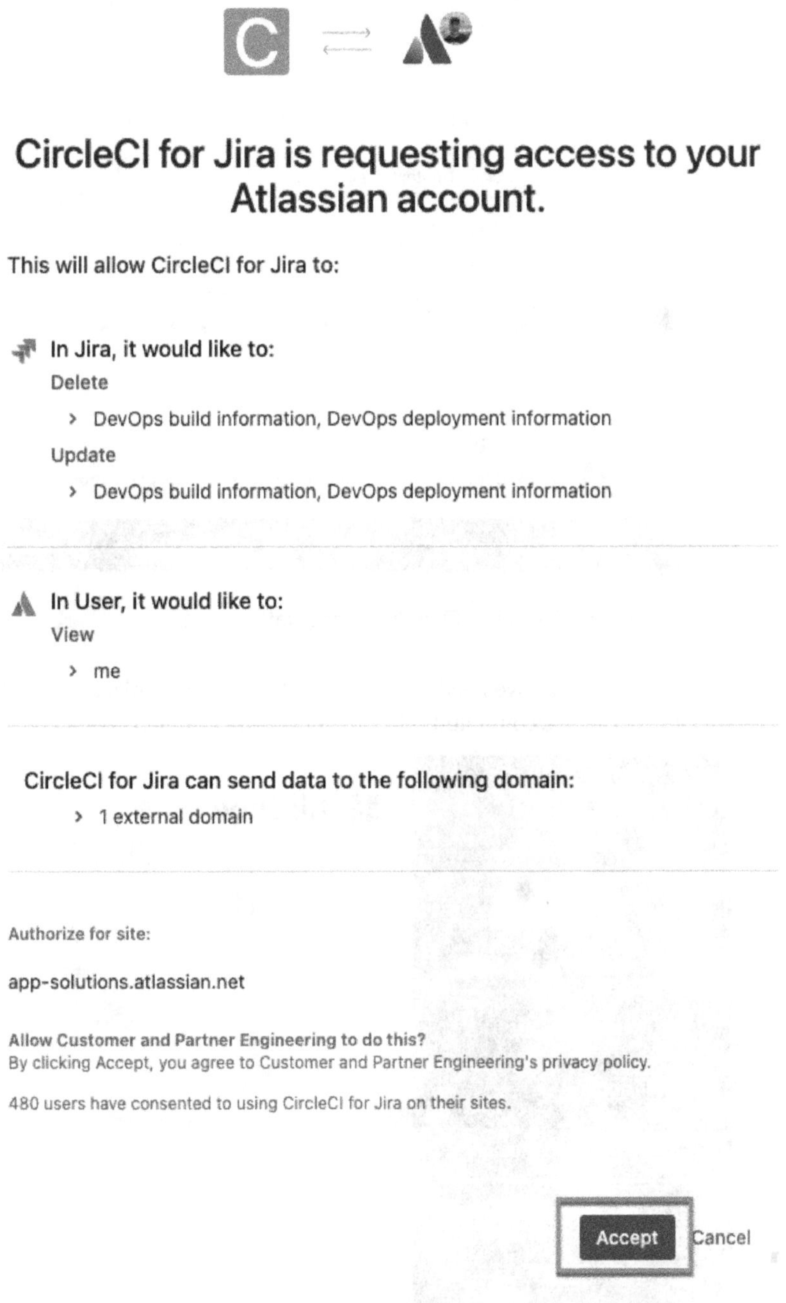

CircleCI for Jira is requesting access to your Atlassian account.

This will allow CircleCI for Jira to:

In Jira, it would like to:

Delete

> DevOps build information, DevOps deployment information

Update

> DevOps build information, DevOps deployment information

In User, it would like to:

View

> me

CircleCI for Jira can send data to the following domain:

> 1 external domain

Authorize for site:

app-solutions.atlassian.net

Allow Customer and Partner Engineering to do this?
By clicking Accept, you agree to Customer and Partner Engineering's privacy policy.

480 users have consented to using CircleCI for Jira on their sites.

Figure 4.53 – Accepting access request

A **CircleCI for Jira – Configure** page is displayed.

CircleCI for Jira - Configure

Web Trigger URL

https://128227a1-b1f0-4c2b-b7b6-1b32a1afee83.hello.atlassian-d

Paste this URL into your CircleCI Context used by the Jira Orb.

CircleCI Organization ID *

You can find it by navigating to **Organization Settings > Overview** in the CircleCI web app.

Audience

The OIDC token audience. **If left empty, the default value is your Organization ID**.

SUBMIT

Figure 4.54 – CircleCI for Jira configuration page

8. To complete the configuration, we need to obtain the CircleCI organization ID. To do this, switch over to your CircleCI account and select the **Organization Settings** menu option.

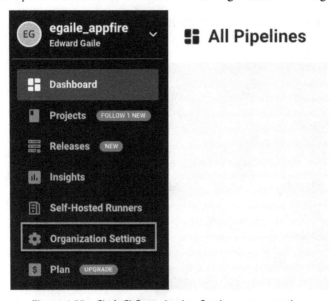

Figure 4.55 – CircleCI Organization Settings menu option

9. An organization **Overview** page will be displayed. An **Organization ID** value is available. Select the copy icon next to the **Organization ID** value.

Figure 4.56 – CircleCI Organization ID value

10. With the organization ID copied to your clipboard, switch back to your Jira application. Paste the organization ID into the **CircleCI Organization ID** field and then select **SUBMIT**.

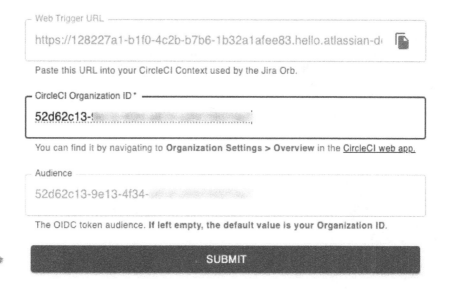

Figure 4.57 – Submitting CircleCI configuration

The CircleCI Jira orb can now be configured with the **Web Trigger URL** value to relay information about your build or deployments. See the following screenshot for where to copy the web trigger ID.

CircleCI for Jira - Configure

Figure 4.58 – Submitting CircleCI configuration

It is beyond the scope of this recipe to fully build out the CircleCI pipeline, but it shows you how to integrate a CircleCI application with Jira. For more information on completing a CircleCI pipeline utilizing the web trigger URL, please refer to the following website: `https://circleci.com/developer/orbs/orb/circleci/jira`.

5

Understanding Bitbucket and Bitbucket Pipelines

Bitbucket is Atlassian's **Source Code Management (SCM)** solution that enables developers to version control their source code. Bitbucket is tightly integrated with Jira and allows developers to collaboratively work on code while also sharing their status with stakeholders. Bitbucket doesn't just store and track changes, it offers a robust deployment system known as **Bitbucket Pipelines**. Using Pipelines, developers can build and deploy their code to various environments with ease. Moreover, **runners** are machines that carry out pipeline executions for build and deployment. Bitbucket Pipelines can use its runners to run a pipeline execution, at a cost of the number of minutes needed for the execution against a monthly budget, or developers can define self-hosted runners that use their own infrastructure.

This chapter has the following recipes – the first recipe will walk you through the basics and subsequent recipes will expand on this foundation:

- Creating a workspace, project, and repository in Bitbucket
- Creating branches in Bitbucket
- Understanding pull requests and merging best practices
- Enabling Bitbucket Pipelines
- Configuring runners in Bitbucket

Technical requirements

You will need the following software:

- Bitbucket Cloud with a repository
- Bitbucket runners

Creating a workspace, project, and repository in Bitbucket

Bitbucket is a very different tool when you compare it against Jira or Confluence. Unlike Jira and Confluence, which require very little configuration to get started, Bitbucket requires some upfront technical configuration before it can be used by a team. This recipe focuses on building your understanding of how Bitbucket is structured and explains the various initial configuration steps required to enable Bitbucket for your team.

How to do it...

When Bitbucket is configured for the first time, you want to first create a **workspace**. Workspaces contain repositories, which will be discussed later in this recipe. When you log in to Bitbucket, you'll see your available workspaces, as shown in the following screenshot:

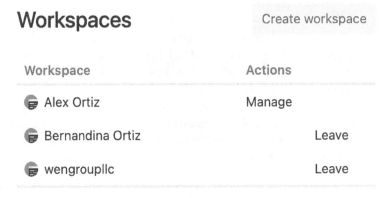

Figure 5.1 – Overview of available workspaces within Bitbucket

> **Important note**
>
> An initial workspace may have already been created when you first signed up for Bitbucket. In that case, you can skip to the next section or, if you want to create a brand-new workspace, follow along. Additionally, make sure you are logged into Bitbucket and that you are an administrator of Bitbucket.

To create a workspace, take the following steps:

1. Click on your profile icon in the top-right corner of Bitbucket.

Figure 5.2 – Select profile icon

2. Click on **All workspaces** under **RECENT WORKSPACES**.

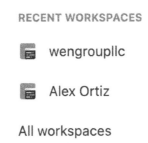

Figure 5.3 – Select All workspaces

3. Click on **Create workspace**, which is located toward the top-right corner of Bitbucket's user interface.

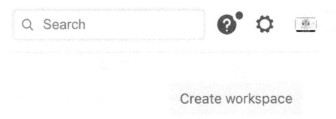

Figure 5.4 – Create workspace button

4. Enter a workspace name.

Figure 5.5 – Provide a name for the new workspace

> **Important note**
>
> This will auto-populate **Workspace ID**. You can optionally change it or leave the default that Atlassian provides. The workspace ID will serve as the URL for the new workspace you are creating.

5.   Update **Workspace ID**.

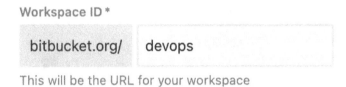

Figure 5.6 – Workspace ID

6.   Determine whether you want the workspace to remain private or whether you want it to be public by checking the checkbox.

✅ **Keep this workspace private**

Figure 5.7 – Determine whether the workspace should be private or public

7.   After you validate the reCAPTCHA, click on the **Create** button:

Figure 5.8 – Create your first workspace

Your workspace is now created, and you will now be able to create your first repository.

There's more...

Once you are in your new workspace, it is time to create your first repository. The repository is where the source code will be stored. You can have many repositories within a single workspace. Let's use the following steps to create a repository:

1.   Click on **Create repository**, located in either the welcome message or under **Recent repositories**.

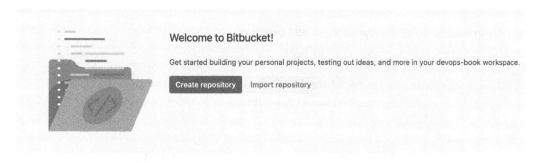

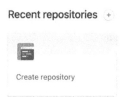

Figure 5.9 – Create your first repository

2.   Bitbucket will redirect you to a new screen where you will first provide a project name. Repositories are grouped into projects, so it is important to provide a project name first:

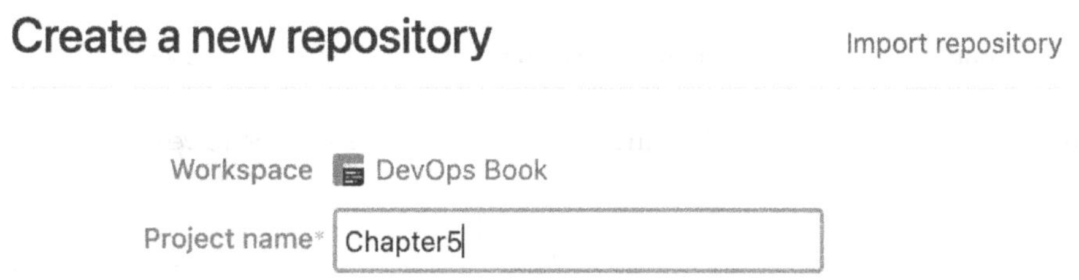

Create a new repository   Import repository

Workspace   DevOps Book

Project name*   Chapter5

Figure 5.10 – Create your first repository and provide a project name

3. Next, provide the name of your repository. This should be a name that is significant to the team that will be using the repository:

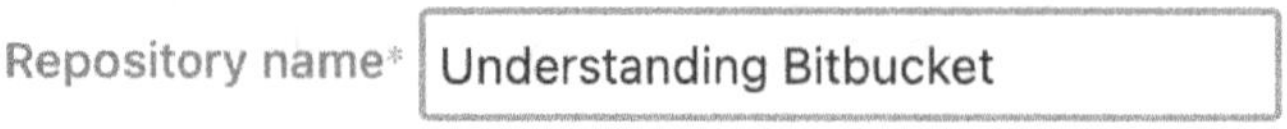

Repository name* Understanding Bitbucket

Figure 5.11 – Provide a repository name

4. Determine whether the repository should be private or public.

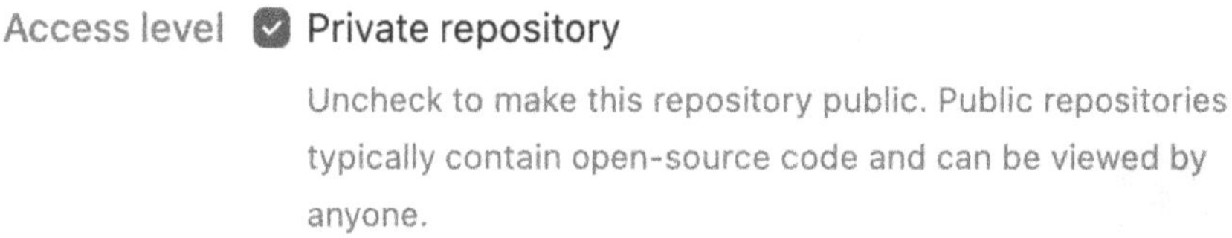

Access level ☑ Private repository

Uncheck to make this repository public. Public repositories typically contain open-source code and can be viewed by anyone.

Figure 5.12 – Determine the access level of the repository

5. Determine whether a README file should be included in the repository, to include it with a template, or to include it with a tutorial, which is recommended for beginners.

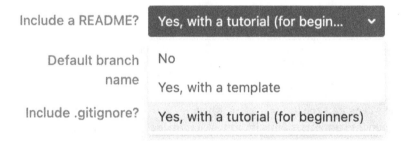

Figure 5.13 – Determine whether a README file is included in the repository

6. Provide the name of your default branch. This is typically main or master.

Figure 5.14 – Name of the default branch

7. Finally, determine whether a `.gitignore` file should be included or not. It is recommended that a `.gitignore` file is included if you have files or directories that need to not be tracked by Git. These files or directories will be ignored and will not be committed to the repository.

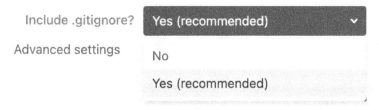

Figure 5.15 – Determine whether a .gitignore file is included or not

8. Click on **Create repository**, located in the bottom-right corner of the screen. Don't worry about configuring any of the advanced settings.

Figure 5.16 – Create repository button

Your first workspace, project, and repository have now been created. In the following recipe, you will learn how to start creating branches in Bitbucket.

Creating branches in Bitbucket

Working off the default branch in Bitbucket is not the best practice as it impacts the main code base, which should be as stable as possible. Instead, you should consider utilizing branches in Bitbucket, and this recipe will cover the basics of branching. There are different techniques when it comes to branching strategies. This recipe will simply cover the mechanics of creating branches.

Getting ready

Before jumping into the mechanics of how to create branches in Bitbucket, let's discuss the different types of branches available. There are different branching strategies in the world of Git. For this example, **Gitflow** will be used to illustrate how branching typically works. Gitflow is well documented by Atlassian, so Gitflow will be used to explain branching strategies in this recipe:

- **Main**: The main branch should be the most stable branch in your repository. This means that the code in the main branch should be the most complete and the best working version of your code. Typically, your main branch is used to create releases and deploy to a production system, so this branch is the most stable.

- **Develop**: This is a very common branch to have when following the Gitflow methodology. The develop branch is the branch your team will integrate their changes into or where they will create feature branches from. The develop branch is initially branched from the main branch.

- **Feature**: Whenever a story is assigned to a developer in Jira, that developer should create a feature branch, which should be based on the develop branch. Feature branches are short-lived and as soon as a developer finishes their code, a pull request should be initiated to merge changes back into the develop branch.

- **Release**: Whenever the Develop branch is ready to be shipped to production, a release branch is created from the latest develop branch. This release branch should be tested and vetted before it is merged into the main and develop branch.

- **Hotfix**: Whenever there are problems with a production system, hotfix branches should be created to address problems with code found in the main branch.

Now that you know what all the different branches are in the Gitflow model, we are going to look at how to create branches within Bitbucket.

How to do it...

Branching in Bitbucket is straightforward but is often not the most popular option. If you are new to using a source control management system, then using a user interface such as that of Bitbucket might be a good option for you:

1. In Bitbucket, enter the repository in which you want to create a branch in. Click on **Repositories** in the navigation bar.

Figure 5.17 – Select Repositories from the menu bar

2. Select the repository you will want to create your branch in.

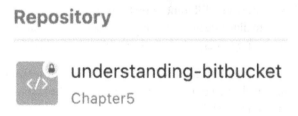

Figure 5.18 – Select the repository of your choice from the available repositories

This will take you to the main repository window.

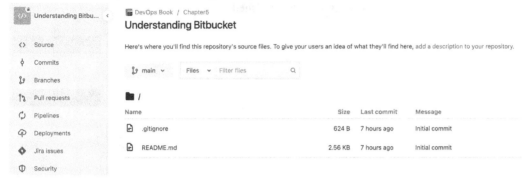

Figure 5.19 – Selected repository

3. Click on **Branches** on the left-side menu.

Figure 5.20 – Select Branches from within the selected repository

4. On the far-right side, toward the top of the screen, click on **Create branch**.

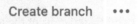

Figure 5.21 – Select Create branch

5. Select the branch type and determine which branch this new branch will be forked from. Give the new branch a name. Click **Create** when finished.

Figure 5.22 – Creating a new branch example

You have successfully created your first branch. The next recipe is going to cover what happens when you finish your code changes and how those changes are shared with the rest of your team.

Understanding pull requests and merging best practices

Tracking branches and changes to your code is only one of the features built into Bitbucket. When you are finished with your code changes, those changes need to be merged back to the branch that is upstream from your current branch. If you are following the Gitflow model, then that means that all of your code changes were done in a dedicated feature branch and now it is time to merge those changes back into the develop branch. The process of doing this merge is called a pull request. A pull request is more than just merging your code changes. The rest of this recipe will explain and walk you through how to use pull requests to maintain high-quality code and minimize the chances that your new code changes will break something.

How to do it...

Once you are done with your changes, it is time to initiate a pull request. This pull request will trigger a chain reaction of events that is covered in the following steps:

1. In Bitbucket, enter the repository in which you want to create a branch. Click on **Repositories** in the navigation bar.

Figure 5.23 – Select Repositories from the menu bar

2. Select the repository you want to create your branch in.

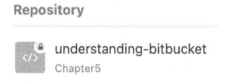

Figure 5.24 – Select the repository of your choice from the available repositories

This will take you to the main repository window.

Figure 5.25 – Selected repository

3.   Click on **Pull requests** on the left-side menu.

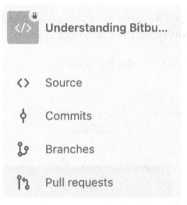

Figure 5.26 – Select Pull requests from within the selected repository

---

Important note

There are many ways to start a pull request. You can initiate it when you are committing your changes. You can initiate it by clicking on the **Create** button in Bitbucket's navigation bar. You can also initiate a pull request from the **Branches** menu. Not to mention that you can initiate a pull request from the command line and from Jira as well. The method described in this recipe is the most comprehensive method for creating a pull request from within Bitbucket.

---

4.   On the far-right side, click on **Create pull request**.

Create pull request

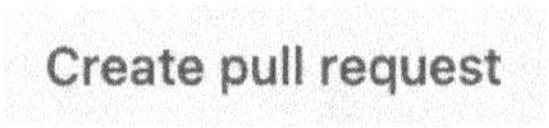

Figure 5.27 – Click on Create pull request

You will be redirected to a screen where you will populate all the details needed to create the pull request.

5.   Start by selecting **Source Branch**. This should be the feature branch you were working on.

Figure 5.28 – Select Source Branch

6.   Select the destination branch that the code changes should be merged into. This should be the **Develop** branch:

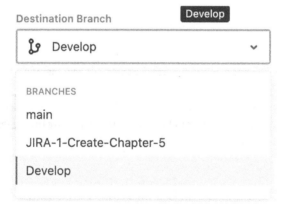

Figure 5.29 – Select the destination branch

7.   Give the pull request a title. This should be anything you want, but as a best practice, this should include the Jira key for the user story/bug that you were working on.

Figure 5.30 – Pull request name

8. Add a description that best describes your changes and anything else that may be helpful to the reviewers.

Figure 5.31 – Provide a description for the pull request

9. The next step is the most critical. Every pull request should be reviewed by someone else on your team. This can be other peers, but preferably, at least one senior-level developer should review the pull request.

Figure 5.32 – Select team members that should review the pull request

10. Optionally, determine whether Bitbucket should delete the branch after it has been merged. This is generally a good idea as Bitbucket will help keep your branches neat by removing unnecessary ones such as those that have been merged already.

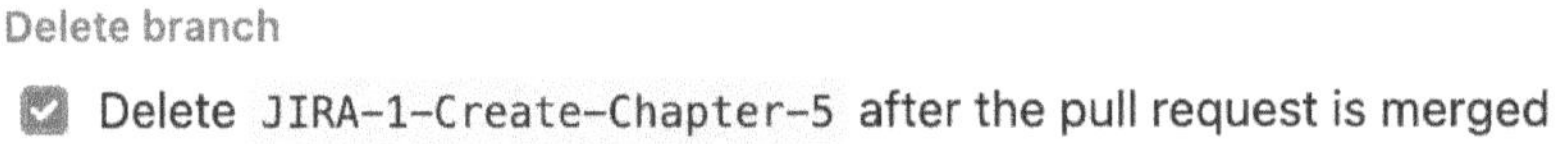

Delete branch

☑ Delete `JIRA-1-Create-Chapter-5` after the pull request is merged

Figure 5.33 – Determine whether the source branch should be deleted after the merge

> **Important note**
>
> Before you create the pull request, review the commits and the files that were impacted as a result of the changes you made. This should serve as a personal review for yourself to ensure that commits or files/changes are not missing from the pull request.

11. Finally, click on **Create pull request** to initiate the pull request.

Figure 5.34 – Provide a description for the pull request

Now that the pull request has officially been made, it is time for someone to review it. Once it has been reviewed, the code can be approved to be merged into the destination branch. In the next recipe, we'll cover how to configure your first pipeline in Bitbucket.

Enabling Bitbucket Pipelines

After all your changes have been approved and merged into a develop branch, they need to be built and deployed. Deploying your changes is important because your source code needs to eventually live in an environment where end users will be able to use the functionalities that you have built. You could manually copy files from your computer to another, but this is not scalable and will most likely not be possible given the distributed world we live in. Instead, you can leverage the power of Bitbucket Pipelines to deploy your code automatically.

How to do it...

Before you can use Bitbucket Pipelines, you will need to tweak a few things:

1. In Bitbucket, enter the repository in which you want to create a branch. Click on **Repositories** in the navigation bar.

Figure 5.35 – Select Repositories from the menu bar

2. Select the repository you want to create your branch in.

Repository

understanding-bitbucket
Chapter5

Figure 5.36 – Select a repository of your choice from the available repositories

This will take you to the main repository window.

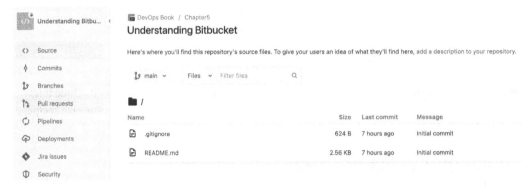

Figure 5.37 – Selected repository

3.   Click on **Pipelines** on the left-side menu.

Figure 5.38 – Select Pipelines from within the selected repository

4. Click on **Create your first pipeline**, which will take you down to the available templates that you can select from.

Build, test and deploy with Pipelines

Easy to set up Integrated CI/CD for Bitbucket Cloud that helps you automate your code from test to production in the Cloud or using your own infrastructure.

Create your first pipeline

Figure 5.39 – Create your first pipeline

5. Select **Starter pipeline**.

Create your first pipeline

Get started with a template   RECOMMENDED

Figure 5.40 – Starter pipeline is recommended

Configuring your first pipeline is an involved process. In the next section, we are going to break down the starter template and explain the different tweaks you can make to the template using Bitbucket's pipeline editor.

There's more...

There are many configurations to configure. This section is going to give you an overview of those changes. In the next chapter, you will learn how to expand on this quite simple pipeline configuration file:

1. The first configuration you can change is the template for your `bitbucket-pipelines.yml` file. In *Step 5* of the previous section, we selected the **Starter** pipeline, but you can select from any of the following templates that Bitbucket provides based on your specific needs. Within your `bitbucket-pipelines.yml` file, make sure you are in edit mode and you will see where you can change the template.

Figure 5.41 – Available pipeline templates

> **Important note**
> Depending on the template that you pick, the information in the template will change.

2. By default, the template you select is going to have recommended steps based on the type of template you selected. If you need to add new steps, you can do so by selecting from the available step types.

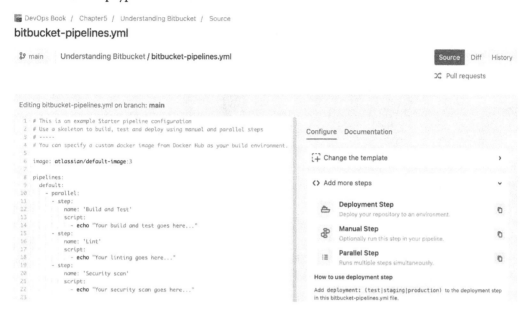

Figure 5.42 – Available step types

3. Select a step that you would like to add and a popup will display, allowing you to copy the information so that you can add it to your template.

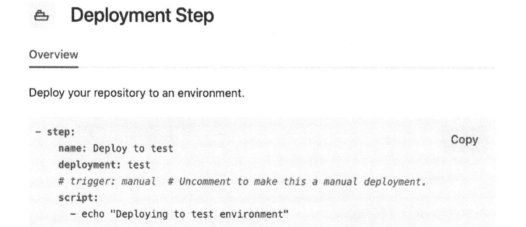

Figure 5.43 – Additional step information to copy

> **Important note**
>
> *Step 3* simply shows how to add a step to your pipeline. Ideally, you want to build steps for building and packaging as well. You may also want to include steps for testing, scanning, delivery, and so on.

4.   Once copied to your clipboard, you can add the additional step back in your template:

```
25          - step:
26              name: Deploy to test
27              deployment: test
28              # trigger: manual  # Uncomment to make this a manual deployment.
29              script:
30                - echo "Deploying to test environment"
31          - step:
32              name: 'Deployment to Staging'
33              deployment: staging
34              script:
35                - echo "Your deployment to staging script goes here..."
36          - step:
37              name: 'Deployment to Production'
38              deployment: production
39              trigger: 'manual'
40              script:
41                - echo "Your deployment to production script goes here..."
42
```

Figure 5.44 – Added deployment step

5. You can also add integrations to your pipeline configuration.

Figure 5.45 – Available pipe integrations

Important note

You can click on **Explore more pipes** to view all the available integrations. There are dozens to choose from based on your specific needs.

6. Finally, you can add variables to `bitbucket-pipelines.yml` to make your pipelines generic, configurable, and reusable.

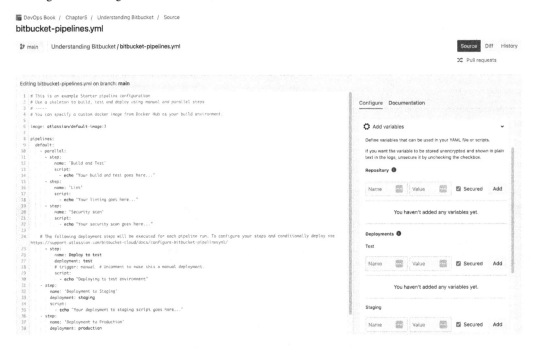

Figure 5.46 – Add variables to the pipeline template

7. Once you have finished modifying your pipeline template, you commit the file back to your repository by clicking on the **Commit** button at the bottom of the screen.

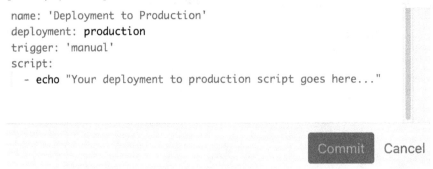

Figure 5.47 – Commit the bitbucket-pipelines.yml file

8. Your pipeline will then execute. After you commit, you will be redirected back to the **Pipelines** menu where you will be able to see the status of your pipeline. I recommend merging the pipeline back to your main or develop branch if needed.

Figure 5.48 – Pipeline status

9. Click on your pipeline execution to obtain details about the latest build run:

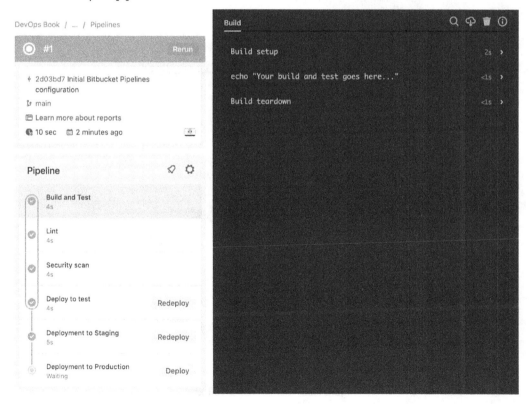

Figure 5.49 – Pipeline run execution details

10. Back in the **Pipelines** menu, you can trigger a pipeline to run by clicking on the **Run pipeline** button.

Figure 5.50 – Force run a pipeline

11. Select your branch and the pipeline you want to run.

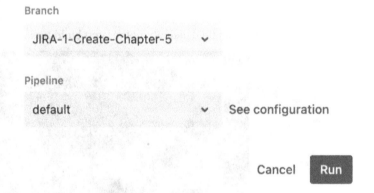

Figure 5.51 – Run Pipeline details

12. You can also schedule pipelines to run at a specific time.

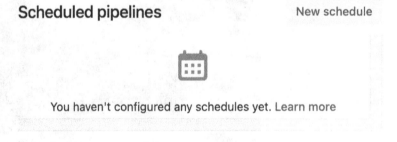

Figure 5.52 – Schedule a pipeline

13. Finally, you can check your usage as Bitbucket pipelines are tracked by the number of minutes it takes to run the pipeline. The amount of time you have each month is determined by your Bitbucket subscription plan.

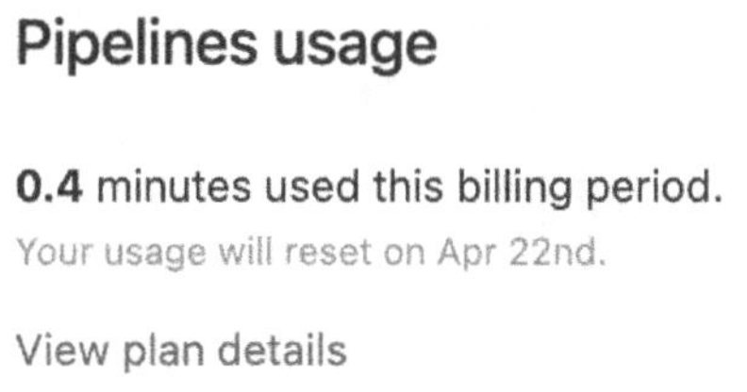

Pipelines usage

0.4 minutes used this billing period.

Your usage will reset on Apr 22nd.

View plan details

Figure 5.53 – View pipeline usage

This was just an introduction to pipelines in Bitbucket. In the next chapter, you will learn how to create more advanced Bitbucket pipelines.

Configuring runners in Bitbucket

With Bitbucket Cloud, you must pay for every build minute. This can quickly add up so, as an alternative, you can utilize self-hosted runners. Runners allow you to save on build minutes, but you need to maintain them in your infrastructure.

How to do it...

Runners will help you keep your Bitbucket costs down, but they do require some initial configurations, which are covered in the following steps:

1. In Bitbucket, enter the repository in which you want to create a branch. Click on **Repositories** in the navigation bar.

Figure 5.54 – Select Repositories from the menu bar

2. Select the repository you want to create your branch in.

Repository

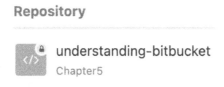
understanding-bitbucket

Chapter5

Figure 5.55 – Select the repository of your choice from the available repositories

This will take you to the main repository window.

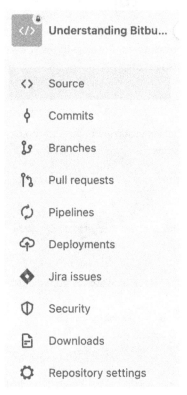

Figure 5.56 – Selected repository

3. Once you are in the repository where you want to add your runner, click on **Repository settings** on the left-hand menu.

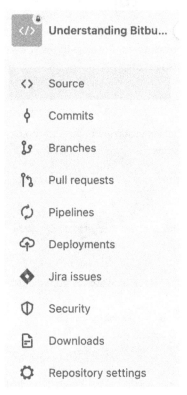

Figure 5.57 – Select Repository settings

> **Important note**
> You can only have up to 100 runners per repository.

4. Once in **Repository settings**, scroll all the way down until you get to the **PIPELINES** section. From there, click on **Runners**.

Figure 5.58 – Select Runners under PIPELINES

5. Click on **Add runner** to add a self-hosted runner.

Runners

Repository runners

No runner installed

Add a self-hosted runner to your build to run pipelines on your own
server. Select **Add runner** to install the runner and add it to the
bitbucket.pipelines.yml configuration. Learn more

Figure 5.59 – Add runner

6. Select the **System and architecture** type for your new runner. There are a few different options, and you should select the one appropriate for the type of runner you are planning on utilizing.

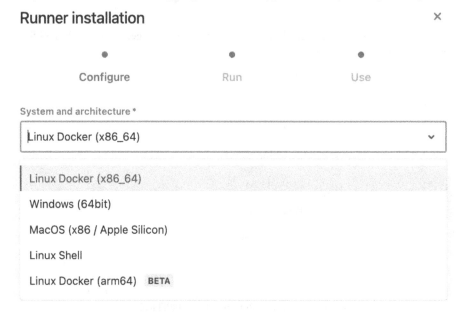

Figure 5.60 – Select the System and architecture type

7. Give your runner a name.

Figure 5.61 – Provide a name for the runner

8. Provide runner labels.

Figure 5.62 – Provide runner labels

> **Important note**
>
> You can provide up to 10 runner labels. Use these labels to help schedule when your runners execute. Labels can only contain lowercase and alphanumerical characters and dots.

9. Click the **Next** button.

Next

Figure 5.63 – Click the Next button to proceed

10. On the next screen, copy the command that is displayed.

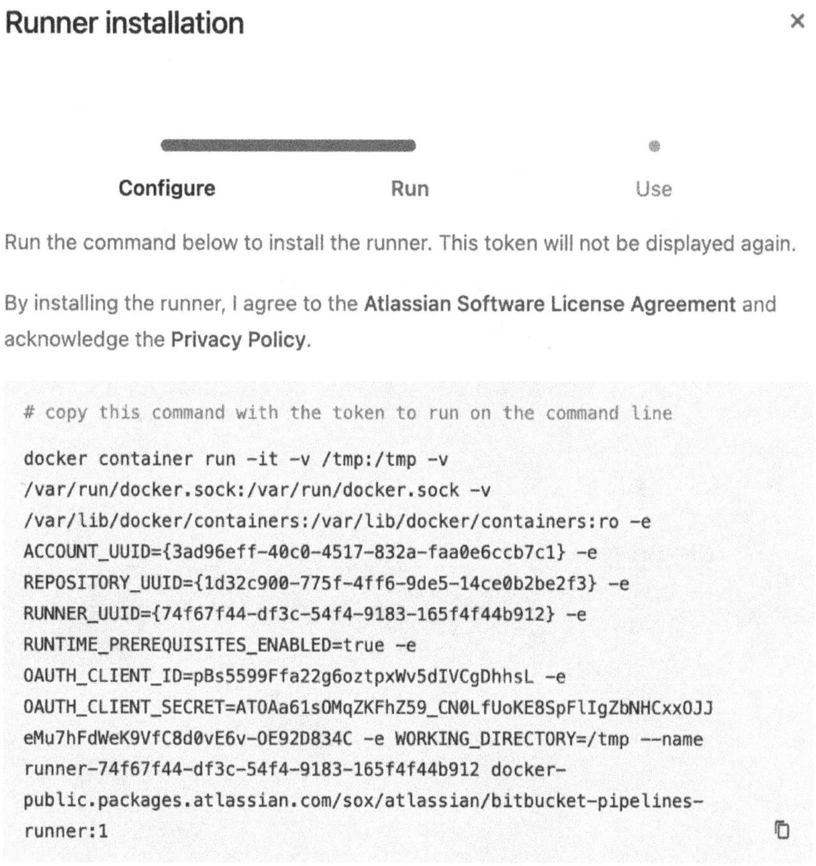

Runner installation   ✕

Configure   **Run**   Use

Run the command below to install the runner. This token will not be displayed again.

By installing the runner, I agree to the **Atlassian Software License Agreement** and acknowledge the **Privacy Policy**.

```
# copy this command with the token to run on the command line

docker container run -it -v /tmp:/tmp -v
/var/run/docker.sock:/var/run/docker.sock -v
/var/lib/docker/containers:/var/lib/docker/containers:ro -e
ACCOUNT_UUID={3ad96eff-40c0-4517-832a-faa0e6ccb7c1} -e
REPOSITORY_UUID={1d32c900-775f-4ff6-9de5-14ce0b2be2f3} -e
RUNNER_UUID={74f67f44-df3c-54f4-9183-165f4f44b912} -e
RUNTIME_PREREQUISITES_ENABLED=true -e
OAUTH_CLIENT_ID=pBs5599Ffa22g6oztpxWv5dIVCgDhhsL -e
OAUTH_CLIENT_SECRET=ATOAa61sOMqZKFhZ59_CN0LfUoKE8SpFlIgZbNHCxxOJJ
eMu7hFdWeK9VfC8d0vE6v-OE92D834C -e WORKING_DIRECTORY=/tmp --name
runner-74f67f44-df3c-54f4-9183-165f4f44b912 docker-
public.packages.atlassian.com/sox/atlassian/bitbucket-pipelines-
runner:1
```

Figure 5.64 – The runner command that is needed to start and configure the runner

> **Important note**
> You will not be able to retrieve the command information later, so make sure you save it in a safe location. You will need this command to install and configure your runner later.

11. Click on the **Next** button.

Figure 5.65 – Click Next to proceed to the next step

12. Copy the labels and add them to your `bitbucket-pipelines.yml` file.

Runner installation   ✕

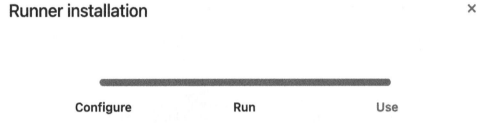

Configure	Run	Use

Copy the labels and add them to **bitbucket-pipelines.yml** in the following format.
Include the `self.hosted` label to unlock extended features for your step. **Learn more**

```
1   # Example
2   pipelines:
3     default:
4       - step:
5           runs-on:
6             - self.hosted
7             - linux
8           script:
```

Your labels

- self.hosted
- linux

Figure 5.66 – Copy labels to include in the YAML file

13. Click on the **Finish** button to finish creating your runner.

Figure 5.67 – Click Finish to create the runner

> **Important note**
>
> Your runner will now be available for your repository but will not be registered until you run the command from *Step 9* of this recipe to start the runner. In the next recipe, you will walk through an example of creating a valid and usable runner.

14. Your newly created runner, with a state of **UNREGISTERED** will be visible in the **Runners** menu within Bitbucket.

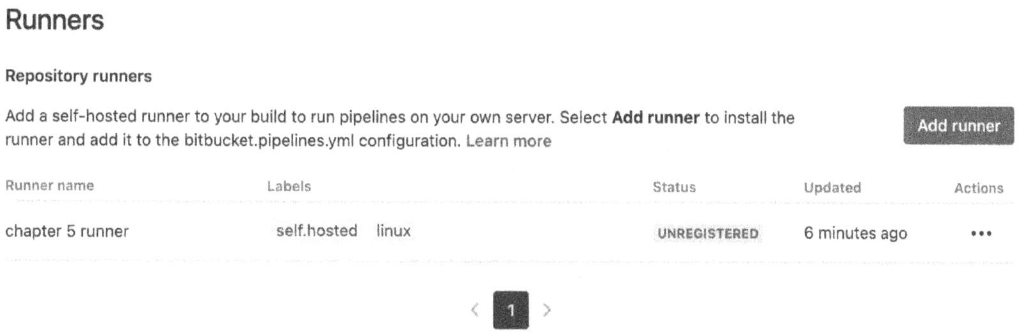

Runners

Repository runners

Add a self-hosted runner to your build to run pipelines on your own server. Select **Add runner** to install the runner and add it to the bitbucket.pipelines.yml configuration. Learn more

Add runner

Runner name	Labels	Status	Updated	Actions
chapter 5 runner	self.hosted linux	UNREGISTERED	6 minutes ago	•••

< 1 >

Figure 5.68 – List of available runners in the repository

Now that you have a runner, you can save on build-minute costs. While you need to host your own runners, it might be an effective way to save a little money if you are consuming a lot of build minutes.

6

Extending and Executing Bitbucket Pipelines

In *Chapter 5*, we started with an introduction to **Bitbucket Cloud**, a cloud-based **source code management** (**SCM**) tool from Atlassian that includes **Bitbucket Pipelines**. Bitbucket Pipelines allows for **continuous integration** and **deployment**, where source code is built, tested, and deployed automatically once a commit is made.

In this chapter, we will start by looking at continuous integration using Bitbucket Pipelines. You will extend your knowledge of Bitbucket Pipelines by adding integrations to third-party tools that perform testing for syntax checking and security. You will also define and configure runners and the agents that perform the execution of Bitbucket Pipelines. Recipes in this chapter include the following:

- Configuring pipeline options
- Conditional execution of pipelines
- Manual execution
- Scheduled execution
- Defining variables
- Defining a runner for a pipeline
- Connecting to Bitbucket Pipes
- Testing steps in Bitbucket Pipelines
- Security steps in Bitbucket Pipelines
- Reporting test results

Let's begin our examination of Bitbucket Pipelines by looking at its structure.

Technical requirements

The recipes in this chapter require the setup of Bitbucket Cloud on a workspace, project, and repository. In addition, **self-hosted runners** require a machine running Linux, macOS, or Windows.

The sample code for this chapter can be found in the `Chapter6` folder of this book's GitHub repository (`https://github.com/PacktPublishing/Atlassian-DevOps-Toolchain-Cookbook/tree/main/Chapter6`).

Configuring pipeline options

This recipe will show you the general structure of the `bitbucket-pipelines.yml` file and the options for general configuration. We created the file that describes the pipeline to execute (`bitbucket-pipelines.yml`) in *Chapter 5*. Let's look at the configuration found in `bitbucket-pipelines.yml` now.

A sample of the possible configurations and the general structure of the `bitbucket-pipelines.yml` file is detailed in the following code snippet:

```
options:
  Global options are here (not required - this section may be absent)
clone:
  Configurations for Git clone behavior go here
definitions:
  Cache and service container definitions here
image:
  Docker image options are here (Chapter 9)
pipeline:
  Pipeline start options are here
    parallel:
      Parallel step options go here
    stage:
      Stage options are here
    step:
      Step options are here
```

This recipe will show you the options for the general configuration of your pipelines. Subsequent recipes will demonstrate the applications, using the options outlined in the preceding code snippet.

How to do it...

We will evaluate the following configuration sections in this recipe:

- Global options

- The `git clone` behavior

- `definitions`

Subsequent recipes will talk about the configuration of the execution options for your pipelines.

Configuring global options

Global options specify behavior for all of the pipelines executed. The global options include the following:

- `docker`

- `max-time`

- `size`

Let's examine these options one by one:

- If you want your pipelines to run Docker commands, specify the `docker` keyword followed by the `true` value, as seen in the following code snippet (we examine this in detail in *Chapter 9*):

```
options:
    docker: true
```

- If you need to allocate more time for steps to execute before timing out, specify the `max-time` keyword with an integer between `1` and `120`. The number specifies the wait time in minutes. The `max-time` option may also be used at the step level to set the timeout for the step. The following code snippet sets the max time to 5 minutes for all steps in the pipeline:

```
options:
    max-time: 5
```

- The `size` option can allocate additional memory to the entire pipeline or an individual step. If you are using runners on Bitbucket Cloud, your options are `1x` and `2x`, while self-hosted Docker runners have options of `1x`, `2x`, `4x`, and `8x`. The following code snippet sets the size of the entire pipeline to double the normal allotment:

```
options:
    size: 2x
```

We have seen how to configure global options relating to running Docker commands, wait times, and available runner memory. Let's now look at options for configuring behavior when performing a `git clone` operation in the pipeline.

The git clone behavior

The clone: section in bitbucket-pipelines.yml controls the copy of the repository through a git clone operation. It can be placed after the options: section to configure the git clone behavior for the entire pipeline or within a step: section to configure the git clone behavior for that step. The options available are outlined in the following list:

- depth: This controls the depth of the clone operation.
- lfs: This allows support for Git **Large File System** (**LFS**) files.
- enabled: This enables or disables the git clone operation.
- skip-ssl-verify: This allows the skipping of the **Secure Sockets Layer** (**SSL**) verification on an individual step.

Let's look at these options in depth one by one:

- You can enable or disable the git clone operations for a specified scope by adding the enabled keyword with true to enable and false to disable. Here is an example that disables cloning for all pipeline steps:

```
clone:
    enabled: false

pipelines:
    default:
        - step:
            script:
                - echo "Cloning not done"
```

- To control the number of commits to include in a git clone operation, use the depth keyword with the value of full to indicate a full clone operation or a positive integer that indicates how many commits to incorporate in the git clone operation. This feature sets up fast checkouts that don't require the full history, especially for large repositories. The following code snippet illustrates a clone operation that clones the last five commits on all pipeline steps:

```
clone:
    depth: 5

pipelines:
    default:
        - step:
            script:
                - ls $BITBUCKET_CLONE_DIR
```

- Use the `lfs` option with the value of `true` to download all LFS files at the start of every step when set globally. If this is placed within a step, the download of all LFS files will start only at the beginning of that step. The following code snippet illustrates the downloading of all LFS files at every step:

```
clone:
    lfs: true

pipelines:
    default:
        - step:
            name: Download LFS
            script: "Cloning and downloading..."
```

- The `skip-ssl-verify` option is available only within the scope of an individual step and only with self-hosted pipeline runners. Setting this option to `true` disables SSL verification for that step, allowing the use of self-signed certificates. This is illustrated in the following code snippet:

```
pipelines:
    default:
        - step:
            runs-on:
                - 'self.hosted'
            clone:
                skip-ssl-verify: true
            script:
                - echo "Using self-signed certificate"
```

We have now seen what options are available for all pipeline steps or an individual pipeline step for the `git clone` operations. Let's now see what resources can be included in the pipeline by configuring the `definitions` section.

Configuring definitions

The `definitions` section describes additional resources available for all pipeline steps. The following list outlines the type of resources available:

- Caches
- Services
- YAML anchors

We examine how services are configured using Docker containers in *Chapter 9*. Caches allow for the temporary storage of build dependencies and their directories. Let's examine the uses of caches:

1. In the `definitions` section, you can define a cache with a file name or directory path. The path can include wildcard characters expressed as a glob pattern. In the following code snippet, we define and use a cache for a Ruby build:

    ```
    definitions:
      caches:
        my-bundler-cache: vendor/bundle

    pipelines:
      default:
        - step:
            caches:
              - my-bundler-cache # Cache is defined above in the
    definitions section
            script:
              - bundle install --path vendor/bundle
              - ruby -e 'print "Building on Ruby"'
    ```

2. The files that should be monitored for changes can be specified in the `caches` definition block as a `key` section with a listing noted under the `files` keyword. Multiple files can be specified using wildcard characters in glob patterns. The `files` location will be specified by the `path` option. The following code snippet shows the implementation of a cache definition for a Ruby build using a node defined by the `key`, `files`, and `path` keyword:

    ```
    definitions:
      caches:
        my-bundler-cache:
          key:
            files:
              - Gemfile.lock
              - "**/*.gemspec" # glob patterns are supported for
    cache key files
          path: vendor/bundle

    pipelines:
      default:
        - step:
            caches:
              - my-bundler-cache # Cache is defined above in the
    definitions section
            script:
              - bundle install --path vendor/bundle
              - ruby -e 'print "Hello, World\n"'
    ```

3. YAML anchors allow for the reuse of blocks of YAML lines. The anchor character (&) defines the reusable block. An alias character (*) serves as the instruction to use the reference. The following code snippet uses YAML anchors to reuse the definition of a step:

```
definitions:
   steps:
      - step: &package
          name: Build and test
          script:
             - mvn package
          artifacts:
             - target/**

pipelines:
   default:
      - step: *package
```

> **Important note**
>
> The names of YAML anchors and aliases cannot contain the following characters: ' [', '] ', ' { ', ' } ', and ', '.

See also

The following links provide more details on caches, especially pre-defined caches to use when building with standard languages and tools:

- `https://support.atlassian.com/bitbucket-cloud/docs/cache-dependencies/#Pre-defined-caches`

- `https://support.atlassian.com/bitbucket-cloud/docs/use-glob-patterns-on-the-pipelines-yaml-file/`

We have seen how to configure options globally on a pipeline. We are now ready to examine the more modular parts of a pipeline's structure.

Conditional execution of pipelines

We turn from specifying the definition options of a pipeline to specifying the execution of a pipeline. This requires us to look at the following sections of a pipeline's structure:

- Pipelines

- Parallel

- Stage

- Step

We will see how these sections are affected in the following use cases:

- Commit against a branch

- Create a pull request

- Creating a tag

Before we examine our use cases, let's visit our structure.

Getting ready

The pipelines section in the bitbucket-pipelines.yml file defines all the available pipeline definitions to build and deploy. It is defined only once in the file and is required.

Within the pipelines section are sections that define pipelines based on the conditional use cases mentioned in the preceding section. A pipeline can be thought of as a series of steps, defined by the step keyword. Pipelines are limited to 100 steps.

The step details at a minimum the commands needed to run the step in a Docker container defined as the build environment. These commands are contained in the required script section. In addition, other options can be defined in the step section to further define the step's behavior. An example of a simple one-command step is shown in the following code snippet:

```
pipelines:
  default:
    - step:
        script:
          - echo "Running a command"
```

Steps are normally run sequentially. If you want to run multiple steps in parallel, group the steps together using the parallel keyword. A common use case is to allow testing steps to occur in parallel. An example is shown in the following code snippet:

```
pipelines:
  default:
    - step:              # non-parallel build step
        script:
          - ./build.sh
    - parallel:          # these 2 steps will run in parallel
        steps:
          - step:
              script:
                - ./integ-tests.sh --batch 1
          - step:
              script:
                - ./integ-tests.sh --batch 2
```

```
        - step:           # non-parallel deploy step
            script:
              - ./deploy.sh
```

A stage is a grouping of steps so that a certain function is performed. The stage section is defined by the `stage` keyword and includes its steps under the `steps` keyword. Stages help define the grouped steps of a distinct phase such as build, test, packaging, and publishing. One use of stages may be in deployment where they can show you which part of a deployment failed and allow for the use of common environments and environment variables. An example of a stage is shown in the following code snippet:

```
pipelines:
    default:
      - stage:
          name: build/test
          steps:
            - step:
                name: Build step
                script:
                  - sh ./build-app.sh
            - step:
                name: Unit-test
                script:
                  - sh ./run-unit-tests.sh
```

The `default` section defines the steps to be run on a pipeline on every push to the repository unless the push occurs on a defined branch-specific pipeline or the push is based on a `git tag` operation. Let's revisit an example of our one-step pipeline and see that because of the `default` keyword, it will always run:

```
pipelines:
    default:
      - step:
          script:
            - echo "Running a command"
```

Now that we understand the basic structure of our pipeline through the `pipelines`, `parallel`, `stage`, and `step` sections, let's visit the use cases where we want the conditional execution of the pipeline steps.

How to do it...

The *Getting ready* section of this recipe showed us the pipeline structure inside of `bitbucket-pipelines.yml`. The `pipelines`, `parallel`, `stage`, and `step` sections allowed for the unconditional execution of pipeline steps.

We now want to run different pipeline steps based on the following use cases:

- Commit to a specific branch
- Creation of a pull request
- Creation of a specific tag

Bitbucket Pipelines allows these conditional use cases through additional sections, denoted with specific keywords. Let's examine these keywords one by one:

- The branches keyword allows the specification of the specific branches against which pipeline steps are to be performed. Branches can be identified by their name or grouped together using glob patterns. In this case, the default keyword identifies those steps to be performed against branches not defined in the branches section. The following code snippet illustrates a pipeline with different steps for any push to the main branch and any feature branch:

```
pipelines:
  default:
    - step:
        script:
          - echo "This script runs on all branches that don't
have any specific pipeline assigned in 'branches'."
  branches:
    main:
      - step:
          script:
            - echo "This script runs only on commit to the main
branch."
    feature/*:
      - step:
          script:
            - echo "This script runs only on commit to branches
with names that match the feature/* pattern."
```

- Pull-request-specific pipelines are defined based on the working branch and the pull-requests keyword. These define the specific steps to run when the pull request to the destination branch is created. Working branches can be combined into similar branch categories using glob patterns. Executing the pull-requests pipeline merges the destination branch into the working branch before running. If the merge fails, the pipeline execution stops. Pipelines for pull requests to feature and hotfix branches are illustrated in the following code snippet:

```
pipelines:
  pull-requests:
    feature/*:
      - step:
          name: Build for pull request to feature branch
```

```
      script:
        - echo "feature branch PR!"
    hotfix/*:
      - step:
          name: Build for pull request to hotfix branch
          script:
            - echo "hotfix PR!"
    '**':
      - step:
          name: Build for all pull requests to other branches
          script:
            - echo "all other non-feature, non-hotfix pull
  request!"
```

- To set up a tag-specific pipeline, use the `tags` keyword to denote the section of steps to run when `git tag` matches the pattern defined in the `tags` section. Note that the glob patterns can be applied for the search parameters to broaden the tag search. The following code snippet runs pipeline operations depending on the tag matched in the search:

```
pipelines:
  tags:
    '*-FirstTag':
      - step:
          name: Build for *-FirstTag tags
          script:
            - echo "First tag!"
    '*-SecondTag':
      - step:
          name: Build for *-SecondTag tags
          script:
            - echo "Second tag!"
    '*-ThirdTag':
      - step:
          name: Build for *-ThirdTag tags
          script:
            - echo "Third tag!"
```

We've now seen how to set up pipelines to conditionally run based on a branch, if a pull request is created, and if a tag is created.

See also

The following links add more details to the sections we have discussed in this recipe:

- https://support.atlassian.com/bitbucket-cloud/docs/step-options/
- https://support.atlassian.com/bitbucket-cloud/docs/stage-options/
- https://support.atlassian.com/bitbucket-cloud/docs/parallel-step-options/

So far, we have talked about pipeline executions that are automatically run when a commit or pull request operation occurs. Can we run pipelines manually? We can! Let's examine the means to do so in the next recipe.

Manual execution

Although normally, pipelines are executed automatically on commit or pull requests, it is possible to manually run pipelines. Not only can they be used to rerun automated pipelines, but you can also run pipelines that can only be executed manually.

In addition, we can also specify that a single step in a pipeline should be executed manually. The execution of a pipeline will pause until action is completed by the user.

Let's look at defining manual-only pipelines.

Getting ready

You can set up pipelines that are intended to be only run manually. These pipelines are in their own section, denoted with the `custom` keyword. Each pipeline has a string that describes the name, as seen in the Bitbucket UI and its steps. The following code snippet describes two manual pipelines and an automated branch pipeline:

```
pipelines:
  custom: # Pipelines that are triggered manually
    manual-sonar: # The name that is displayed in the list in the
Bitbucket Cloud GUI
      - step:
          script:
            - echo "triggering for Sonar!"
    deployment-to-prod: # Another display name
      - step:
          script:
            - echo "triggering for manual deployments to prod!"
  branches:  # Pipelines that run automatically on a commit to a
branch
```

```
    staging:
      - step:
          script:
            - echo "Auto execute for push to staging branch."
```

A manual pipeline can also contain variables that are set or updated when that manual pipeline is run. Variables can be described with the `variables` keyword and can be described using the following properties.

- `name`: The variable's name (this is required)

- `default`: The default value for the variable

- `allowed-values`: A list of allowed values

- `description`: A summary of the variable's purpose and settings

The following code snippet shows a manual pipeline with its variables:

```
pipelines:
  custom: # Pipelines that are triggered manually
    us-build: # The name that is displayed in the list in the
Bitbucket Cloud GUI
      - variables:
          - name: IAMRole
            default: "admin"            # optionally provide a default
variable value
            description: "AWS user role"
          - name: AWSRegion
            default: "us-east-1"
            allowed-values:             # optionally restrict variable
values
              - "us-east-1"
              - "us-west-2"
      - step:
          script:
            - echo "$IAMRole manually triggered for a build for
$AWSRegion"
```

We've now defined pipelines that can be run manually. Let's now examine how to run these pipelines, as well as the normally automated pipelines.

How to do it...

Manually executing pipelines is done at the Bitbucket GUI. From the repository view, you can trigger a pipeline run in three places. Let's look at these options.

Running from the Pipelines view

Run the following steps to manually execute a pipeline from the **Pipelines** view:

1. In the repository sidebar, select the **Pipelines** option.

Figure 6.1 – Selecting the Pipelines view

2. On the **Pipelines** screen, click the **Run pipeline** button.

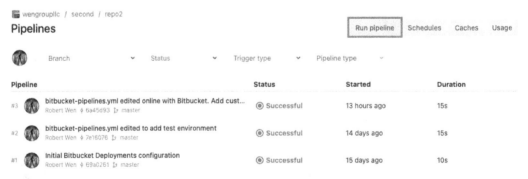

Figure 6.2 – Click the Run pipeline button

3. In the window that appears, select the branch and the pipeline to run. Click **Run**.

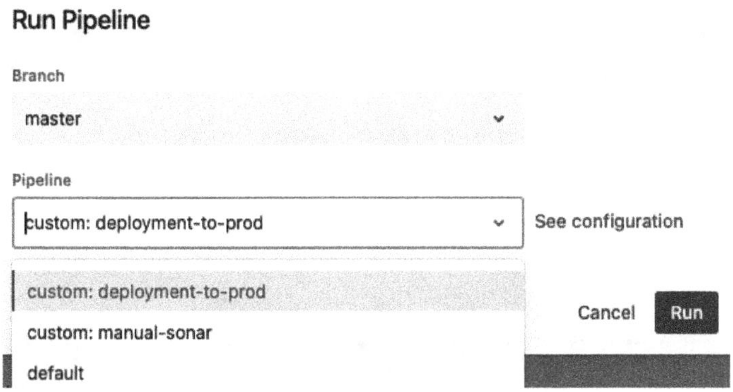
Run Pipeline
Branch
master
Pipeline
custom: deployment-to-prod
See configuration
custom: deployment-to-prod
custom: manual-sonar
default
Cancel
Run

Figure 6.3 – Run Pipeline

You have now manually run a pipeline from the **Pipelines** view.

Running from the Commits view

You can also manually run a pipeline from the **Commits** view. Follow these steps when in the Bitbucket repository:

1. Select **Commits** in the repository sidebar.

repo2
Source
Commits
Branches
Pull requests
Pipelines
Deployments
Jira issues
Security
Downloads
Repository settings

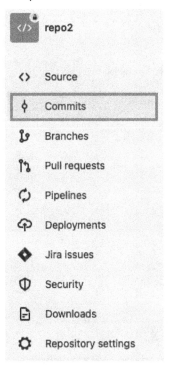

Figure 6.4 – Selecting the Commits view

2.   In the **Commits** screen, select a commit by clicking on its hash.

Figure 6.5 – Selecting a commit

3.   In the **Details** sidebar on the right, select **Run pipeline**.

Figure 6.6 – Select Run pipeline

4. In the window, select the pipeline to run and click **Run**.

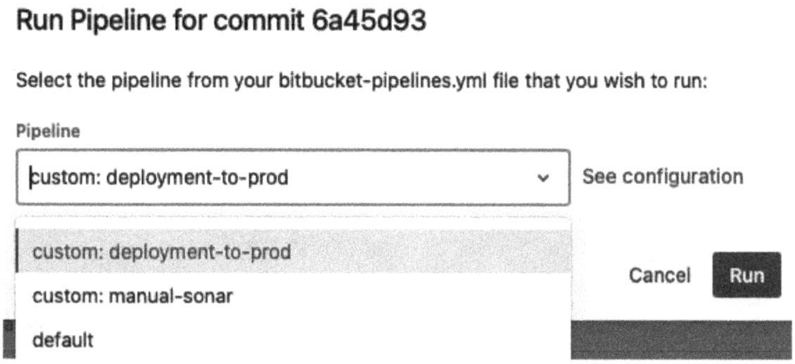

Run Pipeline for commit 6a45d93
Select the pipeline from your bitbucket-pipelines.yml file that you wish to run:
Pipeline
custom: deployment-to-prod
See configuration
custom: deployment-to-prod
custom: manual-sonar
default
Cancel
Run

Figure 6.7 – Run pipeline for commit

You have now manually run a pipeline against a commit.

Run pipeline from the Branches view

The final place to manually run a pipeline is the **Branches** view. Let's see how that's done:

1. In the repository, select **Branches** from the repository sidebar.

repo2
Source
Commits
Branches
Pull requests
Pipelines
Deployments
Jira issues
Security
Downloads
Repository settings

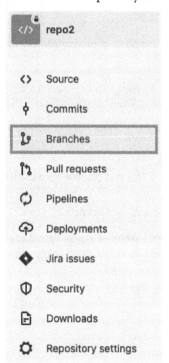

Figure 6.8 – Select the Branches view

2.   On the **Branches** page, find the branch of interest, select the more actions icon (**...**), and select **Run pipeline for a branch**.

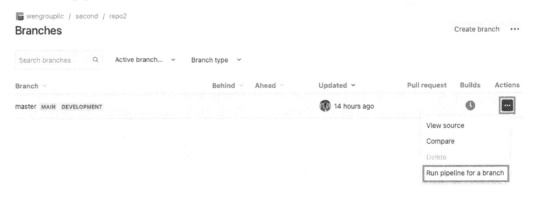

Figure 6.9 – Select Run pipeline for a branch

3.   In the window, select the pipeline to run and click the **Run** button.

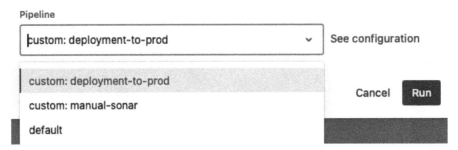

Figure 6.10 – Run pipeline for a branch

You have now manually run a pipeline against a branch.

You can also manually run individual steps of a pipeline. Let's explore how that's done in the next subsection.

There's more...

Any pipeline step that is not the first in the pipeline can be specified as a manual step, requiring an individual to trigger the execution of that step. Let's see how to create this configuration:

1. On the step that is to be run manually, add the `trigger: manual` keyword. This tells Bitbucket Pipelines that this is a manual step. The following code snippet shows a manual step in between two normally run (automatic) steps:

```
- step:
    name: 'First Automated Step'
    script:
        - echo "This step is automated"
- step:
    name: 'Manual Step'
    trigger: manual
    script:
        - echo "This step is manual"
- step:
    name: 'Second Automated Step'
    script:
        - echo "This step is also automated"
```

2. When running the pipeline, click into the pipeline execution from the **Pipelines** view.

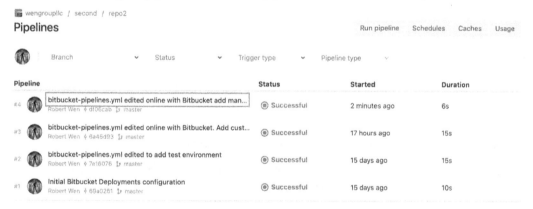

Figure 6.11 – Select pipeline execution

3. In the execution details, to run the manual step, click the **Run** button.

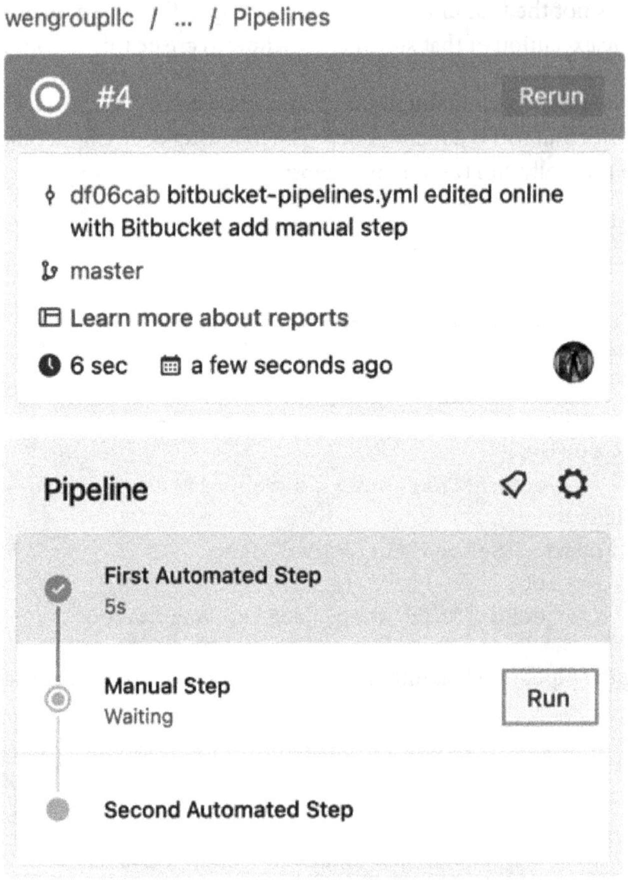

Figure 6.12 – Running the manual step

We have now configured and executed a manual step in our pipeline.

Another desirable feature of pipeline execution is scheduling the execution on a cadence. We will investigate how to do that in our next recipe.

Scheduled execution

Any pipeline defined in the `bitbucket-pipelines.yml` file can be set on a schedule to run on a regular cadence. Let's see how to make the configurations.

How to do it...

Scheduling a pipeline is done on the Bitbucket UI, as seen in the following instructions:

1. In the Bitbucket repository, select the **Pipelines** option in the repository sidebar.

Figure 6.13 – Select the Pipelines view

2.   In the **Pipelines** view, click the **Schedules** button.

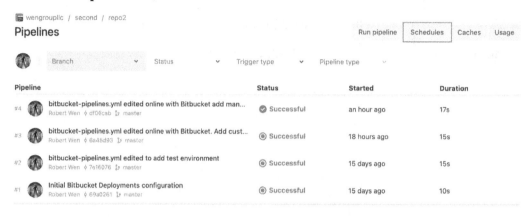

Figure 6.14 – Click the Schedules button

3.   To create a new schedule, click the **New schedule** button in the window that appears.

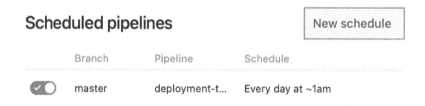

Figure 6.15 – The New schedule button

4.   In the **Create a schedule** window, select a branch, select a pipeline, and select the frequency to run the pipeline. Frequency can be hourly, daily, weekly, or monthly. All times are referred to local time but will be executed on a UTC offset to avoid daylight saving time misconfigurations. Click on the **Create** button when the configuration is complete.

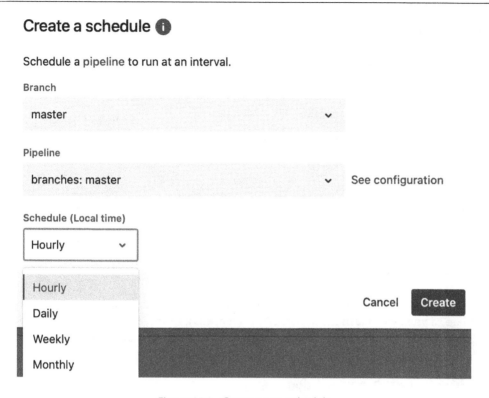

Figure 6.16 – Create a new schedule

5. On an existing schedule, you can click on the slider to disable the schedule. To delete the schedule, hover on the schedule and click on the trash can icon.

Figure 6.17 – Disabling and deleting schedules

We have now seen how to run our pipelines on a regular cadence through scheduling.

In the chapter so far, we have looked at execution from conditional, to manual, to scheduled. We now pivot to looking at what we are executing and enhancements that make our pipelines more potent. The first step is showing how to integrate Bitbucket Pipelines with third-party tools. One way of doing that is using **Pipes**. Let's see how to configure and use Pipes to integrate.

Connecting to Bitbucket Pipes

When we perform continuous integration on our build, we may want to perform testing or security scans to ensure our resulting build has high quality and is secure. Bitbucket Pipelines allows for testing and security scans through the use of integrations. The integrations between Bitbucket Pipelines and an external tool or environment are called Bitbucket Pipes.

We are going to see what pipes are available to Bitbucket Pipelines for integrating with third-party tools and environments. We will also see how to add pipes to your `bitbucket-pipelines.yml` to integrate our pipeline.

How to do it...

The Bitbucket editor for `bitbucket-pipelines.yml` can display the Pipes available for use. The Pipes are then placed in the `script` section of the `bitbucket-pipelines.yml` file, as seen in the following instructions:

1. As we saw in the introduction to Bitbucket Pipelines in *Chapter 5*, when you edit the `bitbucket-pipelines.yml` file in the Bitbucket GUI, it opens a special editor that adds features for Pipes, templates, guidance on steps, and variables. An illustration of the help panel is shown in the following screenshot.

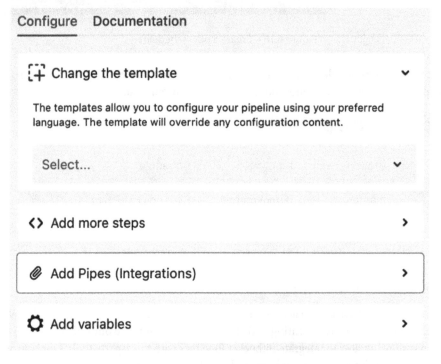

Figure 6.18 – The bitbucket-pipelines.yml editor help panel

2. Expand the **Add Pipes (Integrations)** section to view the available Pipes.

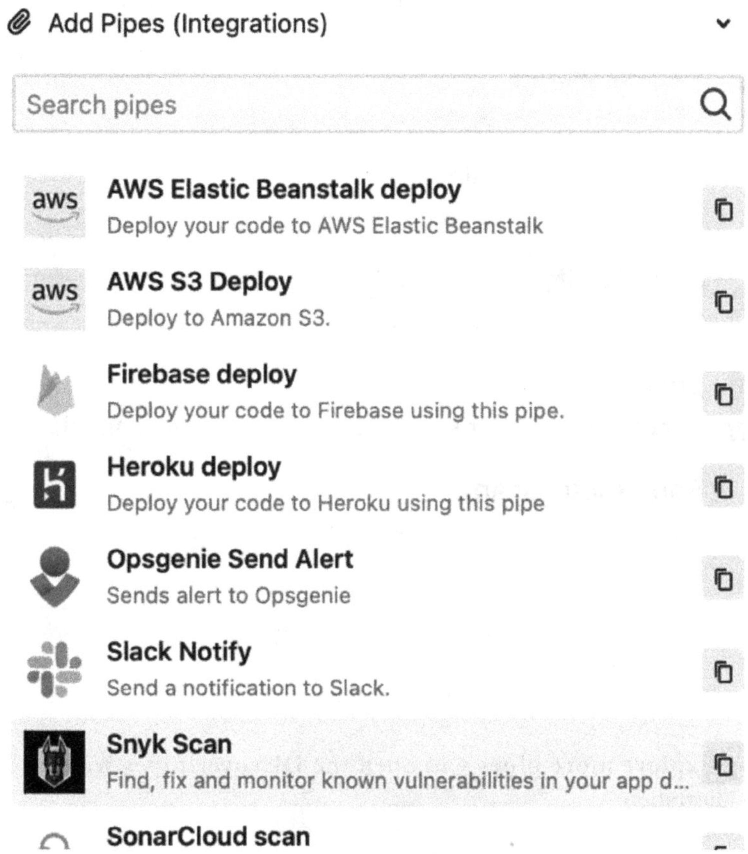

Figure 6.19 – Viewing a selection of Pipes

3. You can search by keyword or view by category when you select **Explore more pipes**, found at the bottom of the Pipes section, as seen in the following illustration.

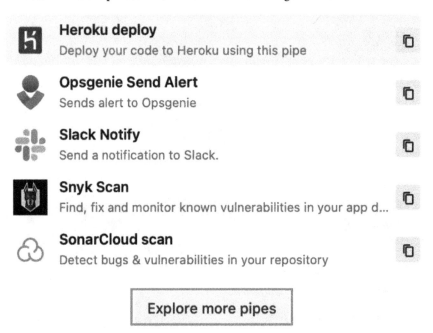

Figure 6.20 – The Explore more pipes button

4. Selecting **Explore more pipes** will open the **Discover pipes** window seen in the following illustration.

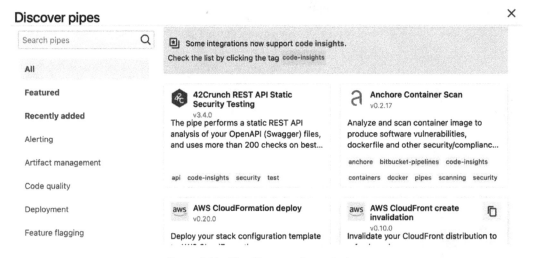

Figure 6.21 – The Discover pipes window

5. Selecting a pipe will open a window for the pipe. This window will contain code snippets to paste into the script section of the step where you want the pipe to run, details about the pipe, and other information. We show an example of this in the following illustration.

Slack Notify

v2.3.0 maintained by Atlassian

Intro   Details   Examples   Support

Sends a custom notification to Slack.

You can configure Slack integration for your repository to get notifications on standard events, such as build failures and deployments. Use this pipe to send your own additional notifications at any point in your pipelines.

YAML Definition

Add the following snippet to the script section of your `bitbucket-pipelines.yml` file:

```
- pipe: atlassian/slack-notify:2.3.0
```
Copy

Figure 6.22 – Pipe window

We will see the application of pipes later in this chapter and *Chapters 8* and *9*.

Another feature that pipelines can use is variables. We started with defining variables in the *Enabling Bitbucket Pipelines* recipe in *Chapter 5*. Let's look at another place to set a variable and how to use variables in your pipeline.

Defining variables

Variables enhance the functionality of your pipelines by allowing you to store values such as names, needed parameters, and sometimes secrets. There are two ways you can define variables, the first of which was referenced in *Chapter 5*. Let's examine these methods.

How to do it...

As we saw in *Chapter 5*, we can define variables from the editor for `bitbucket-pipelines.yml`. Let's see how you can do that:

1.   When editing `bitbucket-pipelines.yml`, select the **Add variables** option.

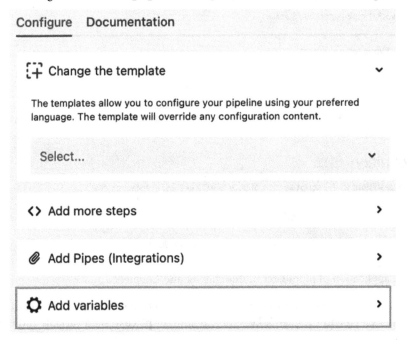

Figure 6.23 – Expand Add variables

2.   You can add variables for the repository, as well as any deployment environments you have defined.

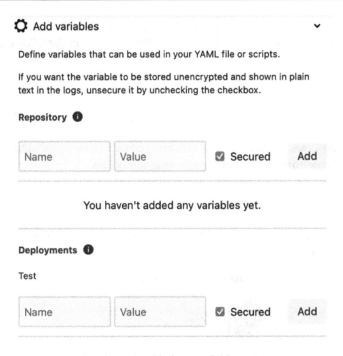

Figure 6.24 – Add variables

3. To add a variable, give it a name, enter its value, and click **Add**. Make sure **Secured** is checked if you need a secured variable for storing passwords or secrets.

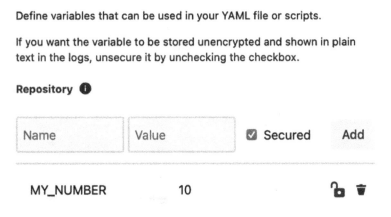

Figure 6.25 – Variable saved

4. To use the variable in `bitbucket-pipelines.yml`, add the variable name to the `script:` section, appended by a dollar sign (`$`) character. This is shown in the following code snippet:

```
- step:
        name: 'Build and Test'
        script:
          - echo "My variable is "$MY_NUMBER
```

We've seen how to add a variable from the `bitbucket-pipelines.yml` editor. Let's see how to add variables from **Repository settings** in the following subsection.

There's more...

Anyone with write access to the repository can create a repository variable. Let's see how to do so from **Repository settings**:

1. At the repository, select **Repository settings** from the repository sidebar.

Figure 6.26 – Select Repository settings

2. In the **PIPELINES** section of the **Repository settings** sidebar, select **Repository variables**.

PIPELINES

Runners

Integrations

SSH Keys

Deployments

Repository variables

OpenID Connect

Settings

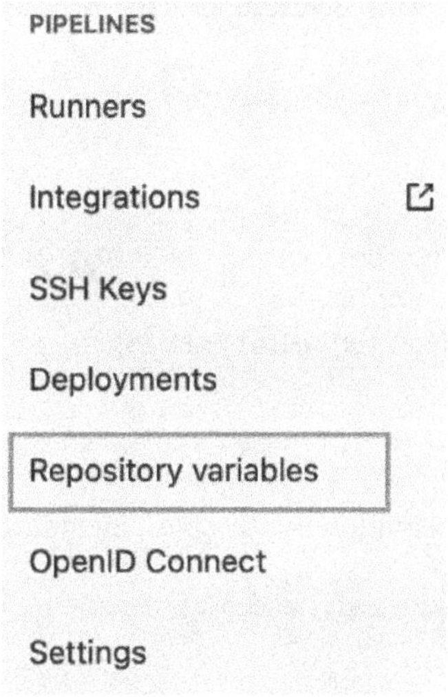

Figure 6.27 – Select Repository variables

3. To add a variable, give it a name, enter its value, and click **Add**. Make sure **Secured** is checked if you need a secured variable for storing passwords or secrets.

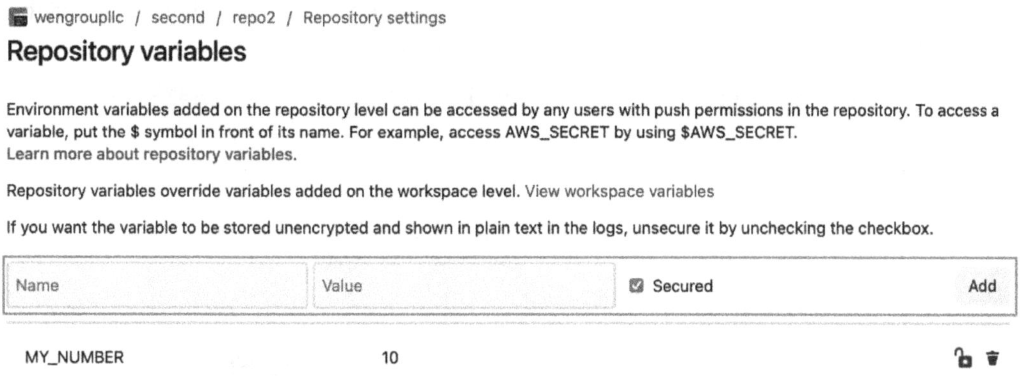

wengroupllc / second / repo2 / Repository settings

Repository variables

Environment variables added on the repository level can be accessed by any users with push permissions in the repository. To access a variable, put the $ symbol in front of its name. For example, access AWS_SECRET by using $AWS_SECRET.
Learn more about repository variables.

Repository variables override variables added on the workspace level. View workspace variables

If you want the variable to be stored unencrypted and shown in plain text in the logs, unsecure it by unchecking the checkbox.

Name	Value	☑ Secured	Add

MY_NUMBER   10

Figure 6.28 – Adding a repository variable

4.   Just as in the previous section, to use the variable in `bitbucket-pipelines.yml`, add the variable name to the `script:` section, appended by a dollar sign ($) character. This is shown in the following code snippet:

```
- step:
        name: 'Build and Test'
        script:
            - echo "My variable is "$MY_NUMBER
```

We've now seen how to add a variable in **Repository settings**.

See also

The following link provides more information on variables, including pre-defined Bitbucket variables:

- `https://support.atlassian.com/bitbucket-cloud/docs/variables-and-secrets/`

We can specify which runners to use when executing a pipeline. Let's explore that now.

Defining a runner for a pipeline

In *Chapter 5*, we saw how to define a self-hosted runner. Bitbucket Pipelines allows the use of self-hosted runners to ensure not only are we using the correct platforms for building, but by specifying our own resources, we can avoid the build time limits of Bitbucket Cloud.

Now that we have runners defined, let's see how to call them in `bitbucket-pipelines.yml`.

How to do it...

Your self-hosted runners can be defined on a step basis in your `bitbucket-pipelines.yml` file. Let's see how that's done:

1.   For a given step, define the runner to use for that step by adding the `runs-on` keyword followed by all the applicable labels. The following code snippet shows such an application:

```
pipelines:
    custom:
        customPipelineWithRunnerStep:
            - step:
                name: First Step
                runs-on:
                    - 'self.hosted'
                    - 'my.label'
```

```
          script:
             - echo "This step will run on a self hosted runner
    that also has the my.label label.";
```

2. Windows-specific runners are used when the `windows` label is added in the `runs-on` section:

```
    - step:
        name: First Step
        runs-on:
          - 'self.hosted'
          - 'windows'
```

3. Mac-specific runners are used when the `macos` label is added in the `runs-on` section:

```
    - step:
        name: First Step
        runs-on:
          - 'self.hosted'
          - 'macos'
```

4. Linux shell runners are used when the `linux.shell` label is added in the `runs-on` section:

```
    - step:
        name: First Step
        runs-on:
          - 'self.hosted'
          - 'linux.shell'
```

5. Linux Docker ARM runners are used when the `linux.arm64` label is added in the `runs-on` section:

```
    - step:
        name: First Step
        runs-on:
          - 'self.hosted'
          - 'linux.arm64'
```

6. If you don't specify a platform label, Bitbucket Pipelines assumes the step should be run by a Linux Docker runner.

7. If all the matching runners are busy, your step may wait until one becomes available. If the runners in your repository do not match any of the labels in the step, the step will fail.

We're now ready to explore some real-life applications of testing steps in a pipeline that are part of continuous integration. Let's see a testing example now.

Testing steps in Bitbucket Pipelines

A key step that may be done on a pipeline for continuous integration typically comes post-build. Testing immediately after the build allows for defects to be found and sorted out, guaranteeing better quality in the code base.

The testing steps in this recipe utilize a category of testing called **static analysis**. With static analysis, the files in the build are scanned to see whether they contain logical errors, security vulnerabilities, or other issues.

Linting is another type of scanning technique that performs tests by evaluating the syntax and structure of the code base. It can also determine whether the code base being evaluated meets specific coding standards.

We will examine different methods of invoking linting and scanning from Bitbucket Pipelines.

How to do it...

Depending on the language, a linting utility may be part of the suite of tools for developing in that language. Let's see how that works with Bitbucket Pipelines:

1. Tests that are part of a language's development platform can be called as a part of the testing step's `script` section. Execute the bundled tests by setting up steps inside the `script` section. The following code snippet features building and testing steps for a Node.js application using **Node Package Manager** (**NPM**) (note that the `image` section denotes a Docker image used as the build environment – we explain this in *Chapter 9*):

    ```
    image: node:10.15.0
    pipelines:
      default:
        - step:
            script:
              - npm install
              - npm test
    ```

2. We can extend our testing by adding linting applications. The following code snippet includes the installation and execution of ESLint, a commonly used linting application for JavaScript in our Node.js environment:

    ```
    image: node:10.15.0
    pipelines:
      default:
        - step:
            name: ESLint
            script:
    ```

```
            - npm install eslint --save-dev
            - eslint --init
    - step:
        name: execute
        script:
            - npm install
            - npm run lint
            - npm test
```

We have now seen examples of running tests when a programming language's development environment contains testing applications. Let's look at integrating a third-party application into our Bitbucket Pipeline in the next subsection.

There's more...

A popular application used for testing is **SonarCloud** from SonarSource. SonarCloud has a battery of tests to measure the security, reliability, and maintainability of a code base. After configuring SonarCloud to Bitbucket Cloud, you can set up test execution in Bitbucket Pipelines in the following manner:

1. SonarCloud Scan utilizes pipes to run the SonarCloud code analyzer from Bitbucket Pipelines. The following code snippet demonstrates the invocation of the pipe in the testing step's `script` section. This sample includes all the optional variables for the pipe. Note that SONAR_TOKEN uses a Bitbucket-secured variable:

```
- step:
    name: SonarCloud
    script:
        - pipe: sonarsource/sonarcloud-scan:2.0.0
            variables:
                SONAR_TOKEN: $SONAR_TOKEN
                EXTRA_ARGS: -Dsonar.projectDescription=\"Project
with sonarcloud-scan pipe\" -Dsonar.eslint.reportPaths=\"report.
json\"
                SONAR_SCANNER_OPTS: -Xmx512m
                DEBUG: "true"
```

2. In addition to SonarCloud Scan, Bitbucket Pipelines can include a call to SonarCloud Quality Gate to perform checks against a defined quality gate before deployment or release. This call is also established as a pipe that can be added to a `script` section. The following code snippet shows the pipe with optional variables:

```
- pipe: sonarsource/sonarcloud-quality-gate:0.1.6
    variables:
        SONAR_TOKEN: $SONAR_TOKEN
        SONAR_QUALITY_GATE_TIMEOUT: 180   # 3 minutes
```

You've now seen how Bitbucket Pipelines connects with SonarCloud analyzers using pipes.

See also

This link includes the steps needed to connect a Bitbucket workspace to a SonarCloud project: `https://docs.sonarsource.com/sonarcloud/getting-started/bitbucket-cloud/`.

A key tenet of DevSecOps is frequently running security scans and testing. Optimally, this should be part of continuous integration. Let's look at an example of including security scanning in Bitbucket Pipelines.

Security steps in Bitbucket Pipelines

As of this writing, **Snyk** is the only security provider that can deeply integrate with Bitbucket Cloud. Adding Snyk as a security provider allows additional security scans into the repository as steps of the development workflow. Part of this workflow includes running security scans in Bitbucket Pipelines. Let's see how that's done.

How to do it...

Snyk uses pipes to integrate with pipelines defined in `bitbucket-pipelines.yml`. Let's look at how to set it up:

1. Add the Snyk pipe to the `script` section of your testing step in the pipeline. Required variables include the Snyk token, the language used (`node`, `ruby`, `composer`, `dotnet`, or `docker`), and the image name if the language is set to `docker`. The following code snippet shows the usage for scanning a Node.js application:

    ```
    script:
      - npm install
      - npm test
      - pipe: snyk/snyk-scan:1.0.1
        variables:
          SNYK_TOKEN: $SNYK_TOKEN
          LANGUAGE: «node»
    ```

2. You can also use the Snyk CLI to run a scan using `snyk test`. Installation of the Snyk CLI can be done through npm. This is illustrated in the following code snippet:

    ```
    script:
      - mvn install
      - npm install -g snyk # binary download also available
      - snyk test --all-projects
    ```

You've now set up Snyk to run security scans as part of a pipeline.

See also

The following link details how to set up Snyk as a security provider for Bitbucket Cloud:

- `https://support.atlassian.com/bitbucket-cloud/docs/add-and-configure-security-with-snyk/`

On some tests, it's possible for Bitbucket to detect the results output and display it. Our last recipe explores this.

Reporting test results

Test results that output in JUnit and Maven Surefire XML formats will be automatically detected by the pipelines. Bitbucket Pipelines then displays any failed test results in a **Tests** tab. Successful executions can be shown in the log view in the **Build** tab. Let's see how to set this up.

Getting ready

When setting up test reporting, make sure the test results are generated in one of the following locations:

- `./**/surefire-reports/**/*.xml`
- `./**/failsafe-reports/**/*.xml`
- `./**/test-results/**/*.xml`
- `./**/test-reports/**/*.xml`
- `./**/TestResults/**/*.xml`

Let's now take a look at how to configure testing results on the pipeline.

How to do it...

Depending on the language, there are a number of different steps to follow to generate test results. Let's look at each method:

1. If you are using the Maven Surefire Plugin in your Maven build job, no special configuration is needed.

2. If you are using PHP and testing with PHPUnit, you should include the `--log-junit` parameter to generate the log output. The following code snippet illustrates the proper command setup:

    ```
    image: php:7.1.1
    pipelines:
      default:
        - step:
    ```

```
        script:
            - apt-get update && apt-get install -y unzip
            - curl -sS https://getcomposer.org/installer | php --
--install-dir=/usr/local/bin --filename=composer
            - composer require phpunit/phpunit
            - vendor/bin/phpunit --log-junit ./test-reports/junit.
xml
```

3. If you are developing with .NET, you can use `JUnitTestLogger` to create the output in JUnit format. The following code snippet illustrates the setup:

```
image: mcr.microsoft.com/dotnet/sdk

pipelines:
  default:
    - step:
        script:
            - dotnet add package JUnitTestLogger --version 1.1.0
            - dotnet test --logger "junit"
```

4. .NET development could also use the `trx2junit` utility to convert the test result files from Visual Studio test result format (`.trx`) to JUnit format (`.xml`). This is shown in the following code snippet:

```
image: mcr.microsoft.com/dotnet/sdk

pipelines:
  default:
    - step:
        script:
            - dotnet tool install -g trx2junit
            - dotnet test --logger 'trx;LogFileName=log.trx'
        after-script:
            - export PATH="$PATH:/root/.dotnet/tools"
            - trx2junit ./TestResults/*.trx
```

You've now seen how to prepare test results so they appear automatically in logs.

7

Leveraging Test Case Management and Security Tools for DevSecOps

In this chapter, we continue on our journey by examining security, a component that is needed to move from a DevOps to **DevSecOps** perspective. This requires additional tools to be integrated into the DevOps toolchain, which is made easier by **Open DevOps**.

In this chapter, we will look at apps from the Atlassian Marketplace that allow for the recording and execution of tests for reference by Jira.

We will also learn how to connect Jira with popular security tools such as Snyk and SonarQube. We'll learn how to track remediations of vulnerabilities discovered in security tools in Jira. Finally, we will learn about the containers used for security testing in Jira and create issues from the vulnerabilities.

After completing this chapter, you should have a good feel for how testing components, whether Jira- or software-related, can be included in an Open DevOps toolchain.

This chapter contains the following recipes:

- Adding test case management to Jira
- Connecting Jira to security tools
- Managing vulnerabilities

Technical requirements

To complete this chapter, you will need the following:

- Jira

- A Snyk account (`https://snyk.io/`)

Adding test case management to Jira

Jira's functionality can be extended into test case management by the utilization of popular marketplace apps such as Xray or Zephyr.

Test case management applications allow Jira users to automate many different types of software testing, including the following:

- Use tests

- Integration tests

- Functional tests

- Acceptance tests

- Performance tests

In this example, we will add the Xray test case management app to Jira.

Getting ready

As indicated in previous chapters, marketplace applications can be installed by product admins. You need to be a Jira product admin to carry out this recipe.

What is Xray?

Xray for Jira is a comprehensive test case management tool that integrates seamlessly with Jira. Xray extends Jira's capabilities, allowing teams to manage their entire testing life cycle directly within the Jira environment.

How to do it...

Xray is a paid application that offers a 30-day free evaluation for users. Follow these instructions to integrate the Xray test case management capabilities into Jira:

1. As the site or product admin, select the **Apps** menu item from the cog icon.

Figure 7.1 – Find new apps

2. This will take you to the **Discover apps and integrations for Jira** page. In the box that you use to search for apps, type `xray` and hit *Enter*. All relevant apps are presented. Select the **Xray Test Management for Jira** app.

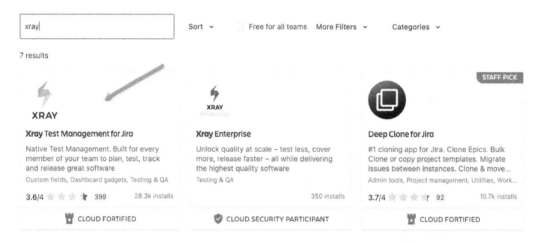

Figure 7.2 – Search for the Xray application

3.   Select **Try it free** to begin the installation process and start the 30-day evaluation.

Figure 7.3 – Get the Xray application

4. The **Add to Jira** pop-up window will present itself. Select the **Start free trial** button to continue the installation process.

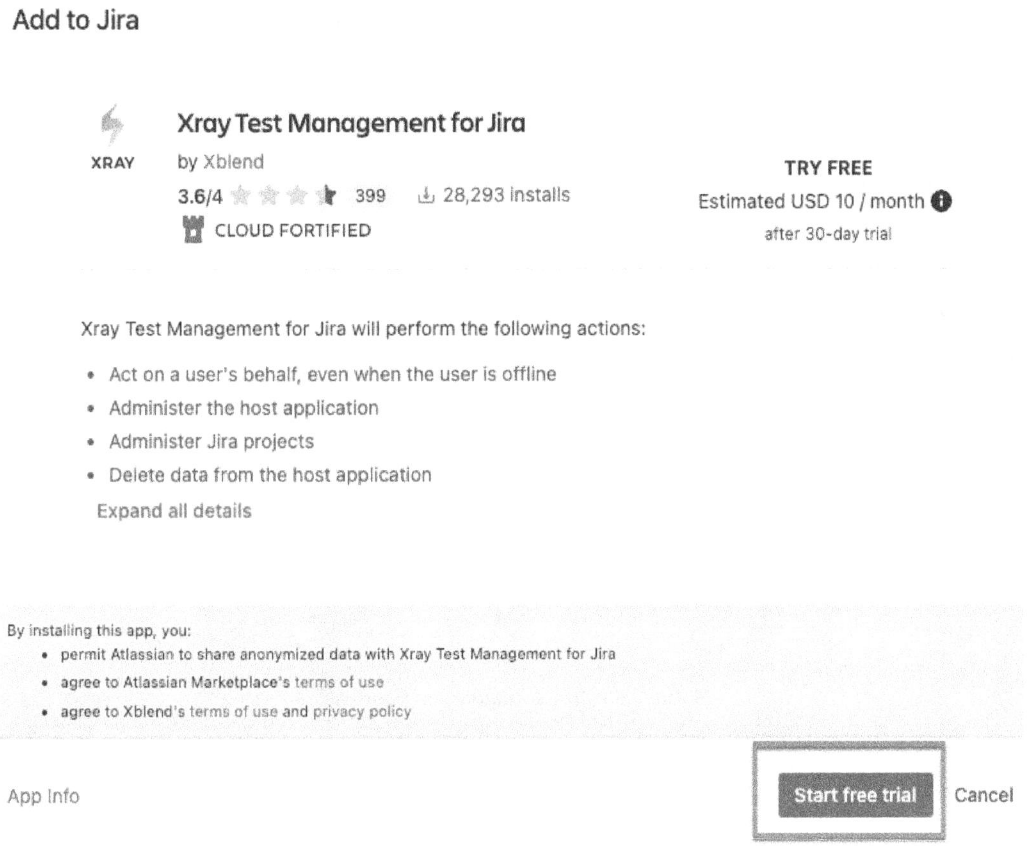

Add to Jira

Xray Test Management for Jira
by Xblend

3.6/4 ★ ★ ★ ★ ★ 399   ⬇ 28,293 installs

🏰 CLOUD FORTIFIED

TRY FREE
Estimated USD 10 / month ⓘ
after 30-day trial

Xray Test Management for Jira will perform the following actions:

• Act on a user's behalf, even when the user is offline
• Administer the host application
• Administer Jira projects
• Delete data from the host application

Expand all details

By installing this app, you:
• permit Atlassian to share anonymized data with Xray Test Management for Jira
• agree to Atlassian Marketplace's terms of use
• agree to Xblend's terms of use and privacy policy

App Info   Start free trial   Cancel

Figure 7.4 – Add the Xray application

5. Once the application is installed and ready, a window will display allowing you to configure the application. Select the **Configure** option.

✓ Xray Test Management for Jira installed   ✕
   successfully

Your app has been added to your instance
and is ready for use.

Get started · Configure · Manage app

Figure 7.5 – Configure the Xray application

6.   The **XRAY** configuration page is displayed. Review the options and make any adjustments necessary. In this example, we will just keep all the default values.

Figure 7.6 – Xray configuration page

7. Next, we need to configure Jira projects with the Xray test case issue types. To do this, select **Xray** from the **Apps** drop-down menu.

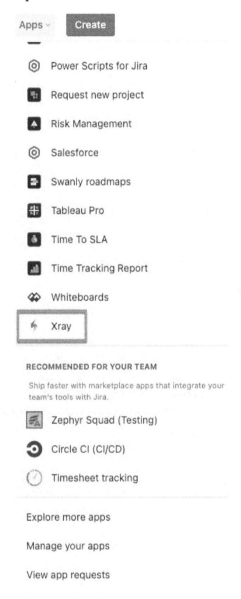

Figure 7.7 – Select Xray from the Apps drop-down menu

8.   The **Get Started with Xray** page is displayed. To configure a Jira project with the test case issue types, select **Configure Project** from the left menu options or from the **Project configuration and organization** panel.

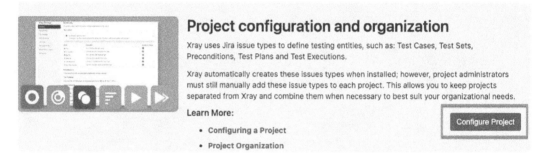

9.   When configuring an existing project, Xray issue types will be added to the specified project. Choose the project to configure from the drop-down menu and click the **Configure** button.

Figure 7.9 – Select project to configure with Xray

10. The Xray project **Summary** page is displayed. The test case issue types are displayed with a red **X** indicating they are not currently present in the project. In the **Xray Issue Types in Project** text box, select the **Add Xray Issue Types** option.

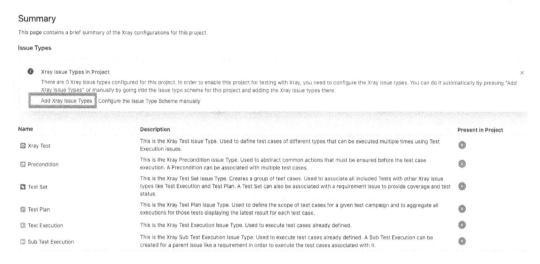

Figure 7.10 – Add Xray issue types to the project

11. A pop-up box will display asking you to confirm that you want to add the issue types to the project's Issue Type Scheme. If the Issue Type Scheme is shared, the new issue types will be added to all other projects associated with the scheme. Select **Yes** to continue; otherwise, you will need to associate the project with a new independent Issue Type Scheme before adding the test case issue types.

Add Xray Issue Types

This project is using an Issue Type Scheme that is shared by other projects. Adding the Xray issue types to this project will change all other projects using the same Issue Type Scheme. Are you sure you want to add the Xray issues types to this project?

Figure 7.11 – Confirm adding Xray issue types

The Xray issue types now have a green checkmark indicating they have been added to the project and are now available.

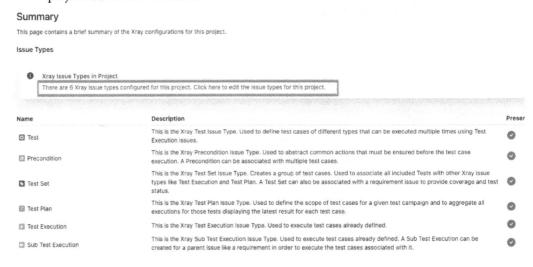

Figure 7.12 – Xray issue types added to project

12. To verify that the preceding steps were completed correctly, select the **Create** button at the top of the **Jira** menu.

Figure 7.13 – Create issue in Jira

13. A new create issue window is presented. In the **Issue type** drop-down menu, you will now see the test issue types available. Go ahead and fill out any required fields to create a new issue with the **Test** issue type.

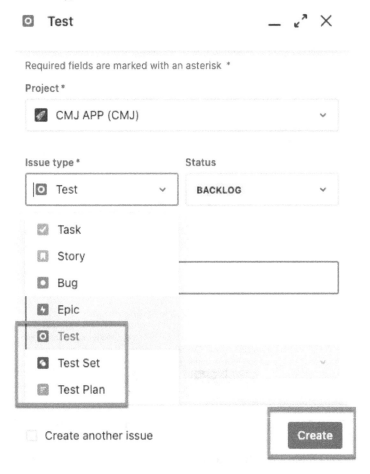

Figure 7.14 – Create a Test issue type

14. Once created, open the new issue. You will now see a **Test details** section panel below the **Description** field. The **Test details** section allows you to add test steps, preconditions, test sets, and test plans, as well as manage test runs.

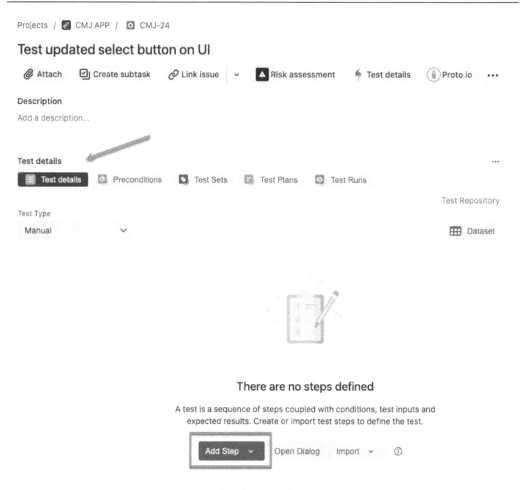

Figure 7.15 – Xray test details available to be added in the issue

For further information on how to utilize the Xray test case functionality, please see the following documentation: `https://docs.getxray.app/display/XRAYCLOUD/About+Xray`

In this recipe, we have successfully installed and configured the Xray test case management app for Jira. Users can now create test case issue types and define the test steps and expected results for these test cases.

Connecting Jira to security tools

Now we will look at the concept of DevSecOps within the Open DevOps toolchain. DevSecOps brings security practices, such as vulnerability scanning, into the CI/CD pipeline. We can achieve DevSecOps by connecting Jira with popular external security tools such as Snyk and SonarQube.

In this recipe, we will connect the Snyk security tool to Jira.

Getting ready

This recipe entails connecting your Jira instance with your Snyk account via the Snyk Security in Jira Cloud marketplace app. In order for this recipe to work correctly, the assumption is that you already have an existing Snyk account (`https://snyk.io/`), you have connected your Snyk account to your source control repository (that is, Bitbucket or GitHub), and you are actively scanning the source control.

You must also be a Jira product admin to execute this recipe.

What is Snyk?

Snyk is an application security scanning tool that specializes in identifying and fixing vulnerabilities in source code. Connecting your Jira and Snyk accounts allows you to manage code vulnerabilities right from Jira.

How to do it...

Let's use the following steps to connect Jira to Synk:

1. As the site or product admin, we need to activate security features for the appropriate Jira project. To do this, go to **Project settings**, then select the **Features** menu option. This will take you to the **Features** page.

Figure 7.16 – Project settings | Features

2. Scroll down the page until you reach the **Development** panel and enable the **Security** section by turning the toggle to green.

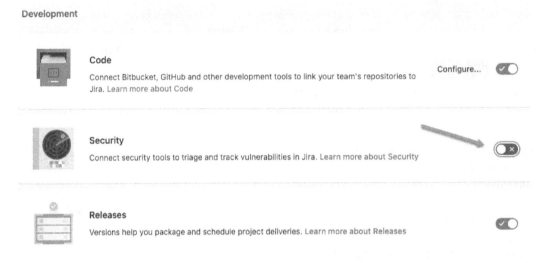

Figure 7.17 – Enable Security

3. Now that we have **Security** enabled for the project, go to the application marketplace and search for the Snyk application. Select the **Snyk Security in Jira Cloud** application.

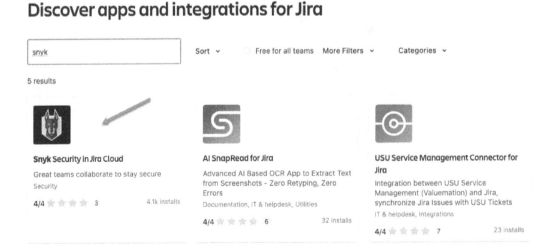

Figure 7.18 – Search for the Snyk Security in Jira Cloud application

4. Select **Get app** to begin the installation process.

Jira / Marketplace apps / Snyk Security in Jira Cloud
Feedback
Snyk Security in Jira Cloud
by Snyk.io
4/4 3 4,142 installs
Get app
FREE
Overview
Support

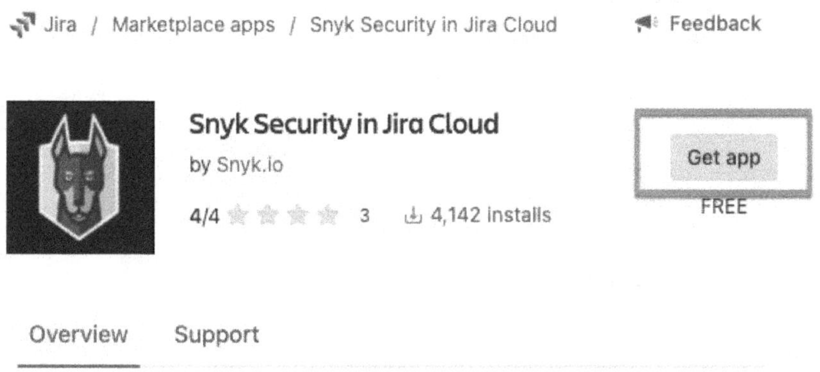

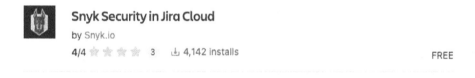

Figure 7.19 – Get the Snyk Security app

5. The **Add to Jira** pop-up window will present itself. Select the **Get it now** button to continue the installation process.

Add to Jira
Snyk Security in Jira Cloud
by Snyk.io
4/4 3 4,142 installs
FREE
Snyk Security in Jira Cloud will perform the following actions:
Delete data from the host application
Write data to the host application
Read data from the host application
By installing this app, you:
permit Atlassian to share anonymized data with Snyk Security in Jira Cloud
agree to Atlassian Marketplace's terms of use
agree to Snyk.io's terms of use and privacy policy
App Info
Get it now
Cancel

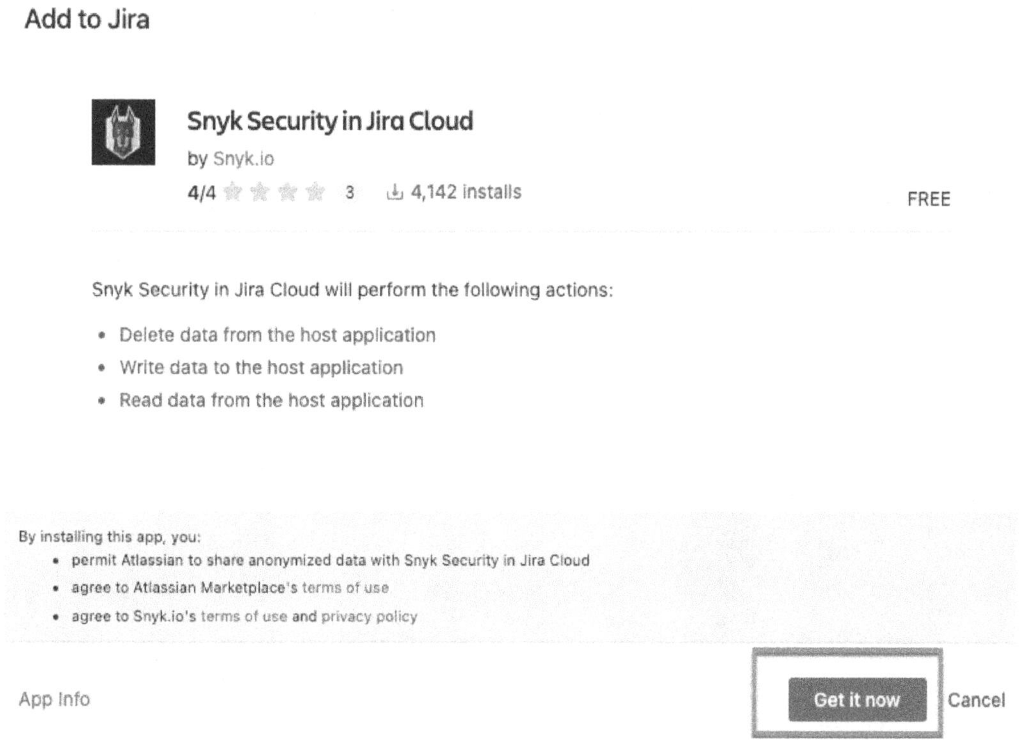

Figure 7.20 – Choose Get it now for Snyk Security

6. Once the application is installed and ready, a window will display allowing you to configure the application. Select the **Get started** option.

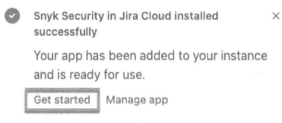

Figure 7.21 – Get started with Snyk Security in Jira

7. The **Snyk configuration** page is displayed. Here, we need to connect to a new or existing Snyk account. For this recipe, we will assume that you already have a Snyk account, so select the **Log in to your account** button.

Figure 7.22 – Snyk configuration

8. A new browser tab will open and display a message asking you to grant Jira access to your Snyk account. Select **Grant app access**.

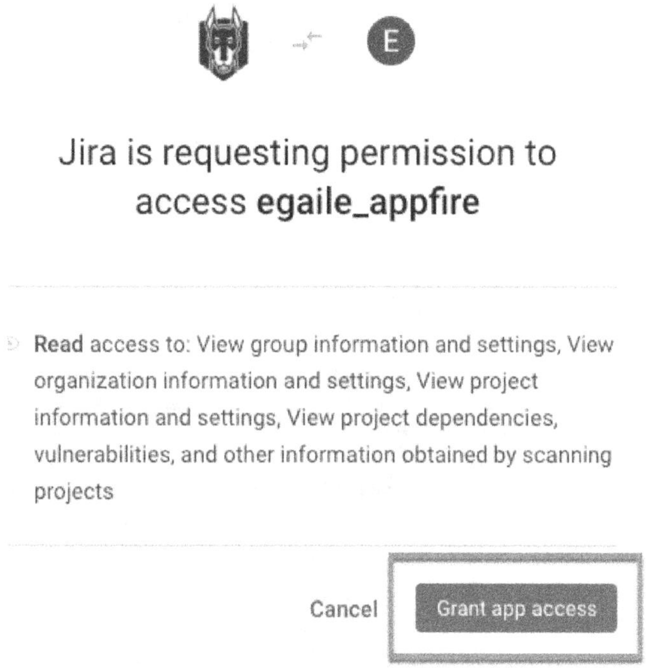

Jira is requesting permission to access egaile_appfire

> **Read** access to: View group information and settings, View organization information and settings, View project information and settings, View project dependencies, vulnerabilities, and other information obtained by scanning projects

Cancel   Grant app access

Figure 7.23 – Grant Jira access to Snyk

9. The new browser window will display a message that your Snyk account has been successfully connected to Jira. Close this browser window and return to the previous browser tab, where you started the connection process.

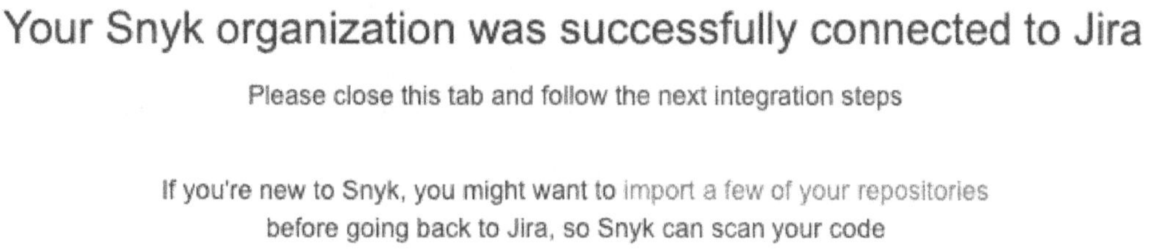

Your Snyk organization was successfully connected to Jira

Please close this tab and follow the next integration steps

If you're new to Snyk, you might want to import a few of your repositories before going back to Jira, so Snyk can scan your code

Figure 7.24 – Snyk access successful

10. The previous browser now displays a message asking you to refresh the updated Snyk configuration. Click on the **Refresh the page** link.

Snyk signup/login process has started

There's an ongoing signup/login process in another tab.
Refresh the page to see an updated view.

Figure 7.25 – Snyk configuration refresh

11. The **Snyk configuration** page now displays your connected Snyk organization. To verify the connection, select the three dots next to the Snyk organization.

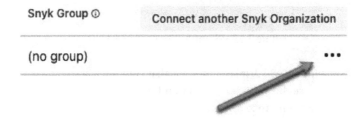

Figure 7.26 – Snyk configuration

12. In the menu option to view the organization and projects in Snyk, select the **Open in Snyk** option.

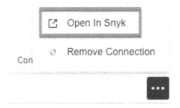

Figure 7.27 – Open to view in Snyk

13. A new browser window will open and display your Snyk Dashboard. Click on the **Projects** menu option on the left side to show the repositories contained in your Snyk organization. This confirms your Jira-to-Snyk integration.

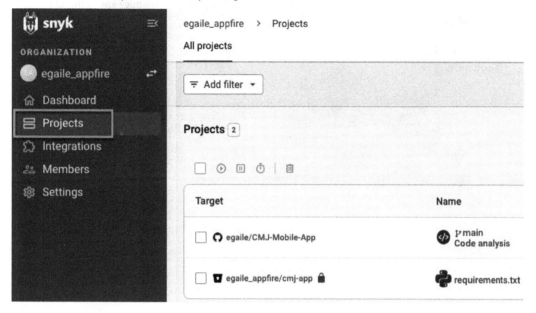

Figure 7.28 – Snyk Projects

Now that we have Snyk and Jira connected, we can look at how to manage vulnerabilities within Jira.

Managing vulnerabilities

Once the appropriate Jira project has been connected to the appropriate Snyk projects, you will be able to see and manage any vulnerabilities resulting from the scheduled or manual code scans. Jira issues can be created from the identified vulnerabilities in order to track and resolve them.

Getting ready

In order to execute this recipe, you will need the following:

- Jira
- Jira administration permissions
- An existing Snyk account (https://snyk.io/)

How to do it...

In the previous recipe, we installed and configured the Snyk for Jira app to connect our Snyk account to Jira. Now we need to add the appropriate Snyk projects to our Jira project to manage any vulnerabilities. Let's use the following steps:

1. To add the Snyk security containers to the Jira project, we need to go to **Project settings | Toolchain**.

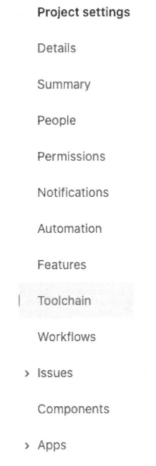

Figure 7.29 – Jira Toolchain project setting

2. The **Toolchain** configuration page is then displayed. Since we installed the Snyk for Jira app, the **Snyk Security in Jira** panel is available. Select **Add security container**.

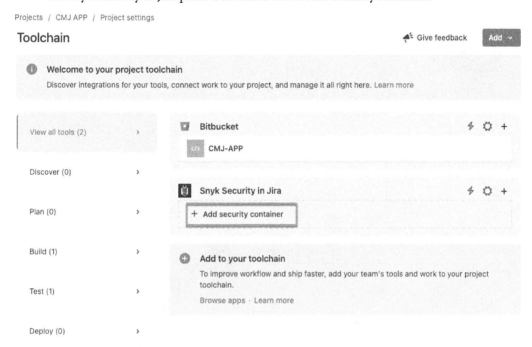

Figure 7.30 – Jira Toolchain configuration page

3. An **Add security container** pop-up window is displayed. Click the drop-down menu to see the available projects in Snyk. Select the appropriate Snyk project and then click the **Add** button.

Figure 7.31 – Add security container

4.   The **Snyk Security in Jira** box in the **Toolchain** window now shows the added security container.

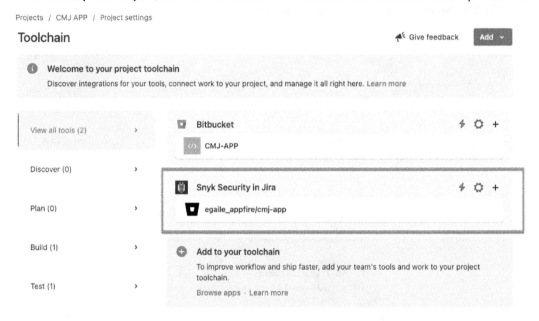

Figure 7.32 – Snyk security container added to project

5. Now that we have the Snyk security container added to the project, we can view the containers for any scanned vulnerabilities. Go back to **Projects** to return to the project summary page. From there, select the **Security** menu option from the **DEVELOPMENT** section.

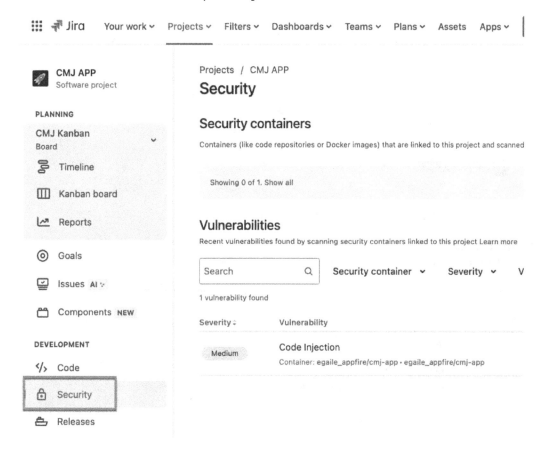

Figure 7.33 – Security menu option

6.   The **Security** page is displayed, showing the attached security containers and any vulnerabilities occurring within that container. In this example, we can see that there is currently a **Medium** code injection vulnerability in the scanned code.

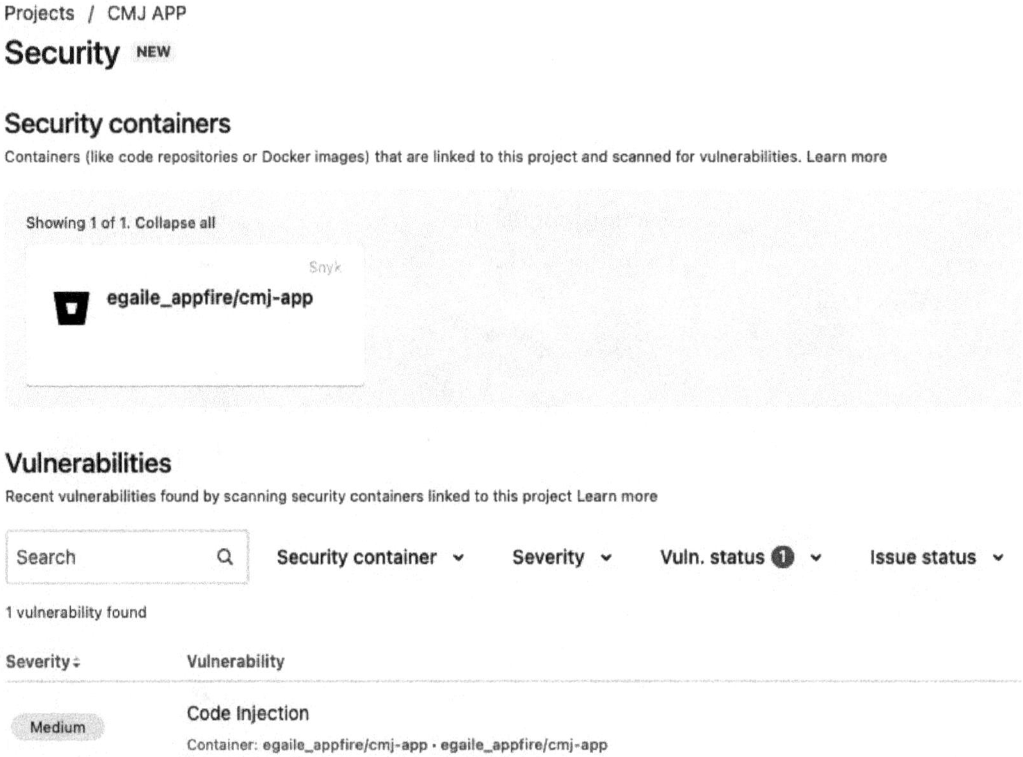

Projects / CMJ APP

Security NEW

Security containers

Containers (like code repositories or Docker images) that are linked to this project and scanned for vulnerabilities. Learn more

Showing 1 of 1. Collapse all

Snyk

egaile_appfire/cmj-app

Vulnerabilities

Recent vulnerabilities found by scanning security containers linked to this project Learn more

| Search Q | Security container ⌄ | Severity ⌄ | Vuln. status ❶ ⌄ | Issue status ⌄ |

1 vulnerability found

Severity ⇅	Vulnerability
Medium	Code Injection Container: egaile_appfire/cmj-app · egaile_appfire/cmj-app

Figure 7.34 – Security containers page

7. If we switch over to the Snyk view of this vulnerability, we can see the same **Code Injection**
    vulnerability in detail.

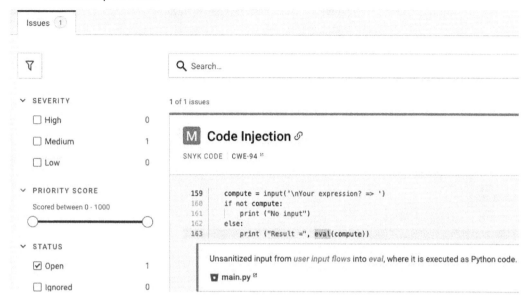

Figure 7.35 – Snyk view of the vulnerability

8. From the Jira **Vulnerabilities** page view, we can create an issue to track and address the
    vulnerability by clicking on the **Create issue** option.

Figure 7.36 – Create Jira issue from vulnerability

9. The **Create issue** window is displayed with the **Description** and **Summary** fields already pre-populated with information about the vulnerability. Select the issue type (that is, **Bug**) and add any other necessary information, then click the **Create** button.

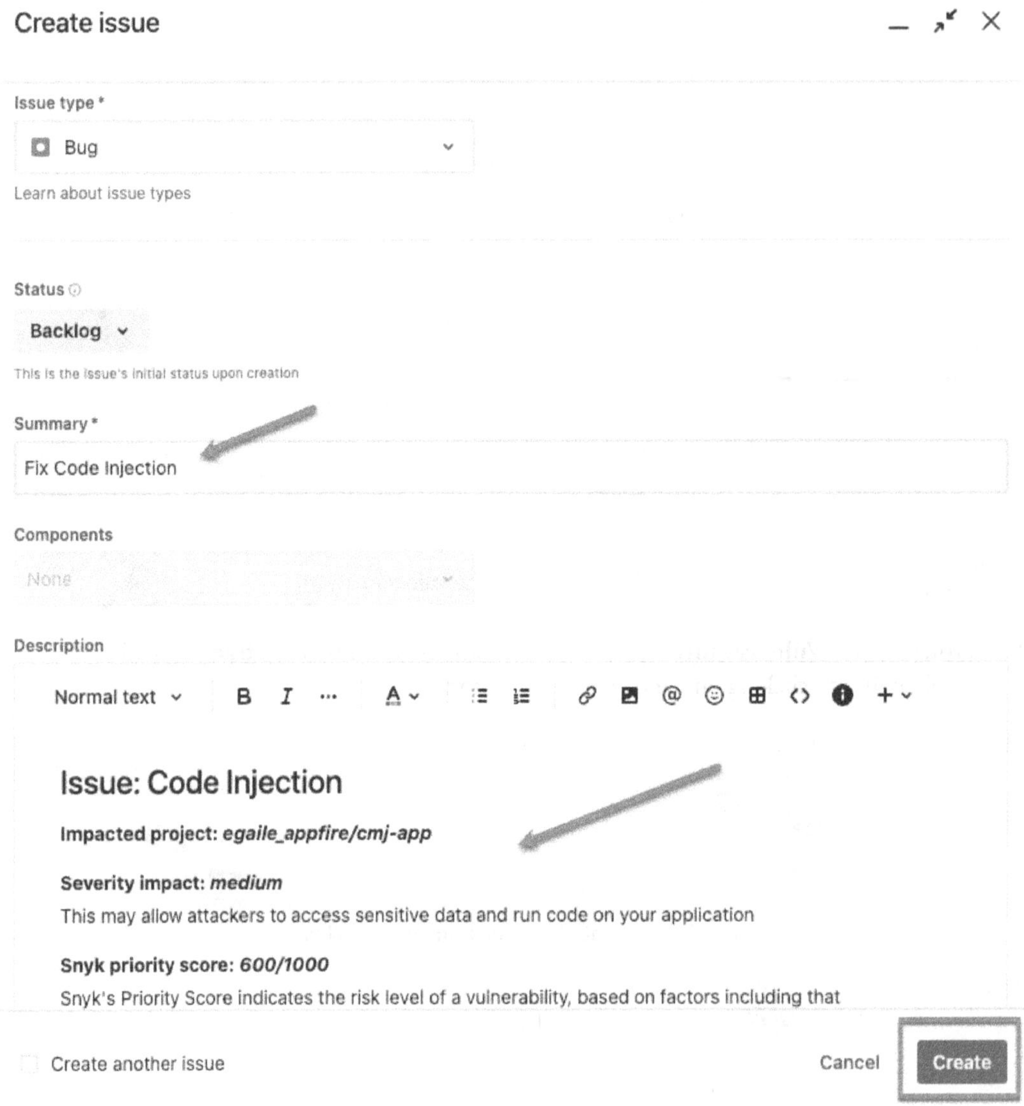

Create issue   — ⬈ ✕

Issue type *

☐ Bug   ⌄

Learn about issue types

Status ⓘ

Backlog ⌄

This is the issue's initial status upon creation

Summary *

Fix Code Injection

Components

None   ⌄

Description

Normal text ⌄   **B** *I* ...   A ⌄   ≡ ≡   ⬀ ▣ @ ☺ ⊞ <> ⓘ + ⌄

Issue: Code Injection

Impacted project: *egaile_appfire/cmj-app*

Severity impact: *medium*
This may allow attackers to access sensitive data and run code on your application

Snyk priority score: *600/1000*
Snyk's Priority Score indicates the risk level of a vulnerability, based on factors including that

☐ Create another issue   Cancel   **Create**

Figure 7.37 – Create issue from vulnerability

10. Once the Jira issue has been created, you will see the Jira issue key available in the **Issues** column of the vulnerability.

Figure 7.38 – Issue linked to vulnerability

11. Click on the Jira issue key to pull up the full issue view.

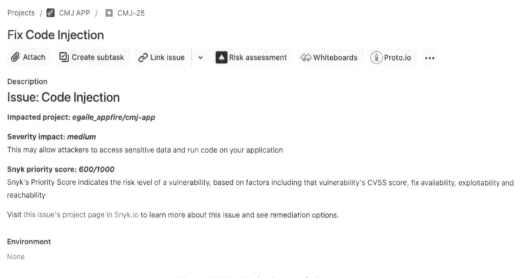

Figure 7.39 – Linked issue full view

In this recipe, we were able to see the real benefits of integrating a code vulnerability app with Jira. We were able to see specific vulnerabilities and simply create a Jira issue to track and correct the vulnerability.

8

Deploying with Bitbucket Pipelines

In the course of understanding Bitbucket Pipelines, we started with the concept of **continuous integration** – that is, we leveraged automation that would perform builds and scanning tests on a commit operation, under the direction of the `bitbucket-pipelines.yml` file.

We are now at the deployment stage, where we must take the build we created previously in Bitbucket Pipelines and install it in one of our environments, from testing to staging, and finally, to production. If we extend Bitbucket Pipelines so that it deploys automatically, we must consider **continuous deployment**.

In this chapter, we'll look at what additions are required in `bitbucket-pipelines.yml` to perform continuous deployment. To do so, we will cover the following recipes:

- Configuring deployments
- Pushing artifacts into the Bitbucket repository
- Pushing artifacts into artifact repository tools
- Deploying artifacts to Bitbucket Downloads
- Deploying artifacts using **Secure Copy Protocol (SCP)**
- Deploying artifacts into AWS S3 buckets
- Deploying artifacts to AWS Lambda
- Deploying artifacts to Google Cloud
- Deploying artifacts to Microsoft Azure
- Using Ansible in the deployment stage
- Using Terraform in the deployment stage

Technical requirements

Because we are still working with Bitbucket Pipelines, we require a subscription to Bitbucket Cloud. This will also be required for recipes where artifacts end up in the repository or Bitbucket Downloads.

Deploying artifacts to cloud-based environments such as AWS, Google Cloud, and Microsoft Azure requires accounts with privileges to the services being used as targets in this chapter's recipes.

The sample code for this chapter can be found in the `Chapter8` folder of this book's GitHub repository `https://github.com/PacktPublishing/Atlassian-DevOps-Toolchain-Cookbook/tree/main/Chapter8`

Now, let's explore how to ready our builds for deployment.

Configuring deployments

While Bitbucket Pipelines can deploy to many different platforms, the characteristics of a deployment are all the same. This can be advantageous when you're deploying to a test environment, where you may want to define some runtime testing or package testing, versus the production environment, where the testing has been exhausted and it's time to release new functionality to the end user.

With this in mind, let's take a look at what the commonalities are for deployments and how to define them uniformly. After, we'll learn how to configure deployments for different environments seen in a typical DevSecOps process.

Getting ready

Before setting up our pipeline, we need to define the environments for deployment. You can use the default environments, which have the following pre-definitions:

- Test
- Staging
- Production

You can also change the names of the environments or configure environment-specific variables for deployment. To set the configurations for your environments, perform the following steps:

1. In Bitbucket, go to the repository of interest and select **Repository settings**.

Figure 8.1 – Selecting Repository Settings

2.   In the **Pipelines** section, select the **Deployments** option.

Figure 8.2 – Selecting Deployments

3.   The list of default environments will appear, divided into **Test**, **Staging**, and **Production**.

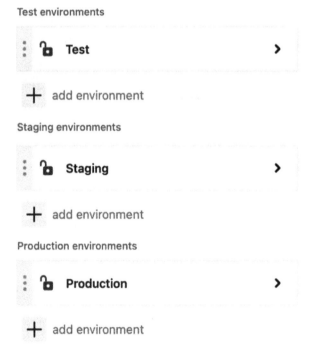

Figure 8.3 – Deployment environments

4. To change an existing environment, click on the band that represents the environment. It will expand, showing options you can use to change the name and a section for environment-specific variables.

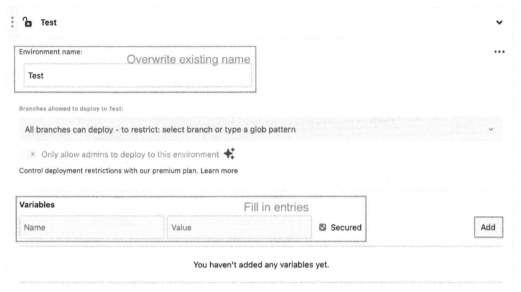

Figure 8.4 – Configuration options for an environment

To change the name, overwrite the name in the **Environment name:** field. To add a variable, fill in the **Name** and **Value** fields. If the variable is meant to contain a secret such as a password or key, make sure that the **Secured** checkbox is selected. Finish by clicking the **Add** button.

5. To add a new environment, go to the section that defines the environment type (**Test**, **Staging**, or **Production**) and select the **add environment** button.

6. Fill in the name of the new environment and select the **Staging environments** checkbox.

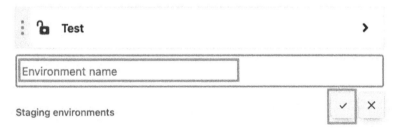

Figure 8.5 – Adding an environment

On the free plan of Bitbucket, you can define 10 environments. On the premium version of Bitbucket, this limit is increased to 100. The premium version of Bitbucket allows you to restrict which branches can deploy to that environment and whether admins alone can perform any deployment on that environment.

Now that we've configured our environments, it's time to configure the deployments.

How to do it...

To define a deployment, we must introduce new keywords and sections to our `bitbucket-pipelines.yml` file. We'll apply these keywords and sections to stages or steps in our pipeline.

Once we've done this, we can define separate deployments, depending on which branch is receiving merges from lower branches. This allows us to continually deploy and test until the final pull request to a main branch, where deployment may happen on the production environment.

Let's look at performing these deployments.

Configuring a deployment in bitbucket-pipelines.yml

The deployment instructions are identified in a stage or step within the `bitbucket-pipelines.yml` file with the `deployment:` keyword, along with the environment that is the target for the deployment step or steps. Let's learn how to use the `deployment:` keyword:

1. The `deployment:` keyword is used within a step to identify the environment that is being deployed. An example is shown in the following code snippet:

    ```
    pipelines:
      default:
        - step:
            name: Deploy to staging
            deployment: staging
            script:
              - python deploy.py staging_1
    ```

    You can define multiple steps for deployment, with each step noting the environment. However, Bitbucket requires that the environment types fall within the following order:

    I.   Test environment.

    II.   Staging environments

    III. Production environments

2. If several steps are required for a deployment, the `deployment:` keyword can be used within the definition of a stage to identify the target environment. An example of this is shown in the following code snippet:

    ```
    pipelines:
      default:
        - stage:
            name: Deploy to Production
            deployment: production
    ```

```
steps:
    - step:
        name: Basic deploy first step
        script:
            - sh ./deploy1.sh
    - step:
        name: Basic deploy second step
        script:
            - sh ./deploy2.sh
```

3. If you want your deployment step to be triggered manually, add the `trigger: manual` keyword within your deployment step. This can be seen in the following code snippet:

```
- step:
    name: Deploy to staging
    deployment: staging
    trigger: manual
    script:
        - python stage_deploy.py staging
```

Once we've defined deployment steps or stages in our `bitbucket-pipelines.yml` file, we can monitor the progress and outcome of our deployments in Bitbucket. Let's examine the process for doing so.

Monitoring deployments

Once you've executed a deployment, you can track its progress or manually execute deployments in the deployment dashboard. Let's examine the features of the deployment dashboard:

1. As we saw in the previous recipe, to reach the deployment dashboard, select **Deployments** from the sidebar at the repository level.

2. The deployment dashboard will appear to the right of the sidebar. It is divided into the environments defined in the `bitbucket-pipelines.yml` file. Within this environment, there will be a card that exhibits the last successful deployment to that environment.

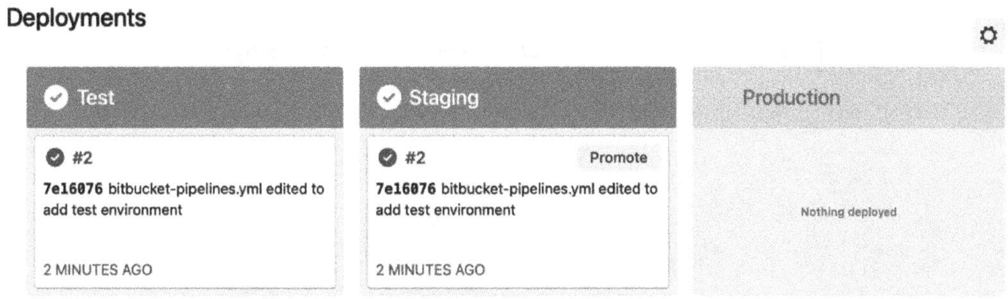

Figure 8.6 – Deployments dashboard

3. Click on the card in an environment to view the details of the deployment for that environment. The details will include the commit that caused the execution of `bitbucket-pipelines.yml`, a note on the differences between the existing version and the commit, and the push to the associated environment, including deployment history. If Jira and Bitbucket are connected, as we discussed in the *Connecting Bitbucket* recipe in *Chapter 1*, you will also see the Jira issue associated with the commit. The following screenshot shows a deployment to the **Test** environment.

Figure 8.7 – Deployment to the Test environment

4. If a deployment is defined to be manually triggered, you will see a **Promote** button on the last successful deployment to the lower environment. Click **Promote** to execute a deployment to the higher environment.

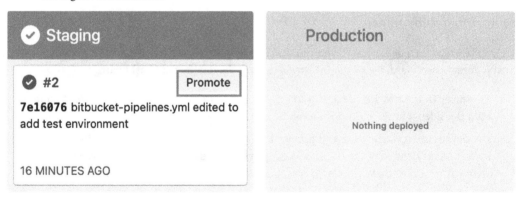

Figure 8.8 – Promoting a manual deployment to production

5. You can redeploy the last successful deployment from the deployment dashboard if a deployment fails. The ability to redeploy requires the following conditions to be met:

    *   An initial deployment step in the pipeline was executed successfully

    *   Deployment permissions are set to allow the redeployment of the step (please note that this is only available in the premium plan of Bitbucket)

    *   Artifacts that are used for deployment can't expire

6. Once a deployment to a specific environment is engaged, any subsequent deployments from other pipelines to that environment will be paused thanks to **concurrency control**, which limits pipeline executions to a single deployment in a given environment. When the original in-progress deployment is complete, you have the following options for paused deployments:

    *   Rerun the pipeline from the beginning

    *   Resume the pipeline from when it was paused

Now that we understand the mechanics of deployment, we can allow deployments to specific environments if a pull request to its associated branch succeeds. Let's look at how that's done.

Deploying from a pull request or branch

As you may recall from the *Conditional execution of pipelines* recipe from *Chapter 6*, you can set up your pipelines to execute different instructions based on which branch is receiving the new commit or if a pull request is being requested. Let's learn how these different conditions can be established:

1. A branch pipeline specifies actions to be performed when a push occurs on a specific branch. You can use the branches: keyword to describe what actions occur on a branch pipeline. These actions are grouped under the branch name. The following code snippet details various actions, including deployment for other branches, when using the default: keyword and separate actions for the staging branch:

```
pipelines:
    default:
        - step:
            script:
                - echo "We do this on all branches except staging"
    branches:
        staging:
            - step:
                deployment: staging
                script:
                    - echo "Done on the staging branch"
                    - python deploy.py
```

2. A pull request pipeline allows you to specify the actions when a pull request is created for a specific branch. These actions are defined on branches underneath the `pull-requests:` keyword. Let's look at actions that are completed when a pull request occurs on the staging branch:

    ```
    pipelines:
        pull-requests:
            staging/*:
                - step:
                    deployment: staging
                    script:
                        - echo "Test and deploy pre-merge"
                        - python deploy.py
    ```

With that, we've learned how to easily test and deploy when changes occur on feature branches. Automated deployment to production is possible when a pull request from staging to the production branch is created.

Now that we've seen how to configure Bitbucket Pipelines for deployment, we will look at examples of how to perform specific deployments depending on the target. First, we'll learn how to deploy into the Bitbucket git repository itself.

Pushing artifacts into the Bitbucket repository

Part of the deployment process may be to take build artifacts and place them in the correct repository. The most convenient repository to store build artifacts may be Bitbucket itself. This is often discouraged because build artifacts typically consume a large amount of storage and may cause performance issues with the underlying git tool. Nevertheless, we offer this recipe if no alternatives exist.

Let's learn how to push build artifacts back into a Bitbucket repository as part of deployment.

Getting ready

Although the recommended way to push back content to the git repository is using HTTP, there may be times when the only way to do this is by using SSH. To do that, you'll need to set up your credentials. This is true if you have branch permissions enabled on your repository or want to set up an automated account for these actions. An important consideration is that configuring these accounts removes limits on what they can access.

Let's learn how to create accounts with different authentication methods.

Using OAuth for authentication

You can use OAuth as the authentication method for accessing the git repository. This involves creating an account and giving that account write access to the main or master branch through branch permissions. Let's learn how to set up an account with OAuth:

1. On your **Workspace** page in Bitbucket, click on the **Settings** cog and select **Workspace settings**.

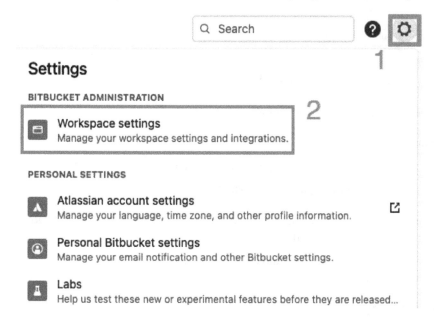

Figure 8.9 – Selecting Workspace settings

2. In the sidebar for **Workspace settings**, find the **APPS AND FEATURES** section and select **OAuth consumers**.

Figure 8.10 – Selecting OAuth consumers

3. On the **OAuth consumers** page, click the **Add consumer** button:

OAuth consumers

Generate your own OAuth consumer key and secret to build your own custom integration with Bitbucket.

Name

You have no consumers configured.

Figure 8.11 – Add consumer

4. On the subsequent page, fill in the following details:

    *   **Name**
    *   Set **Callback URL** to https://bitbucket.org
    *   Make sure **This is a private consumer** is checked

5. For permissions, ensure that both **Read** and **Write** permissions are checked under **Repositories**. Click the **Save** button to save the OAuth consumer:

6. Return to the consumer page and find the new consumer you created. Record the key and secret as secure pipeline variables. Use CLIENT_ID for the key and CLIENT_SECRET for the secret:

OAuth consumers

Generate your own OAuth consumer key and secret to build your own custom integration with Bitbucket.

Name

⌄ OAuth test

   Description

   Key

   Secret

Figure 8.12 – OAuth consumer key and secret

7. Add the following code snippet to the `script` section of the `bitbucket-pipelines.yml` file before the git commands to make the changes and commit to the git repository. We assume that your runner has the `curl` and `jq` utilities installed:

```
- >
  export access_token=$(curl -s -X POST -u "${CLIENT_
ID}:${CLIENT_SECRET}" \
    https://bitbucket.org/site/oauth2/access_token \
    -d grant_type=client_credentials -d scopes="repository"| jq
--raw-output '.access_token')

# Configure git to use the oauth token.
- git remote set-url origin https://x-token-auth:${access_
token}@bitbucket.org/${BITBUCKET_REPO_OWNER}/${BITBUCKET_REPO_
SLUG}
```

With that, you've configured an OAuth consumer and allowed it to make git commits from Bitbucket Pipelines.

We can also set up SSH keys using the Bitbucket UI. Let's examine how to do that.

Creating an SSH key pair for Bitbucket Pipelines

For a given repository, you can create an SSH key pair and save the public key in Bitbucket so that it can be deployed back to the git repository using SSH. Let's see how that's done:

1. On the repository page, select **Repository settings** in the sidebar:

2. In the **Pipelines** section of the **Repository settings** sidebar, select **SSH Keys**:

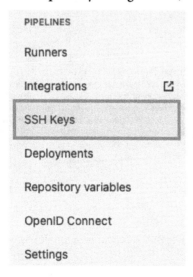

Figure 8.13 – Selecting SSH Keys

3. On the **SSH Keys** page, select **Generate keys**:

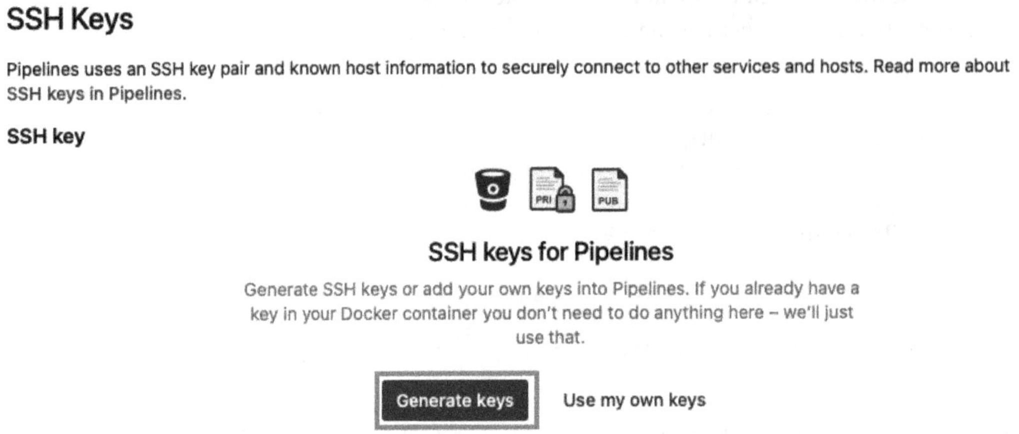

SSH Keys
Pipelines uses an SSH key pair and known host information to securely connect to other services and hosts. Read more about SSH keys in Pipelines.
SSH key
PRI
PUB
SSH keys for Pipelines
Generate SSH keys or add your own keys into Pipelines. If you already have a key in your Docker container you don't need to do anything here – we'll just use that.
Generate keys
Use my own keys

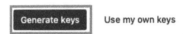

Figure 8.14 – Generating keys

4. Copy the resulting public key:

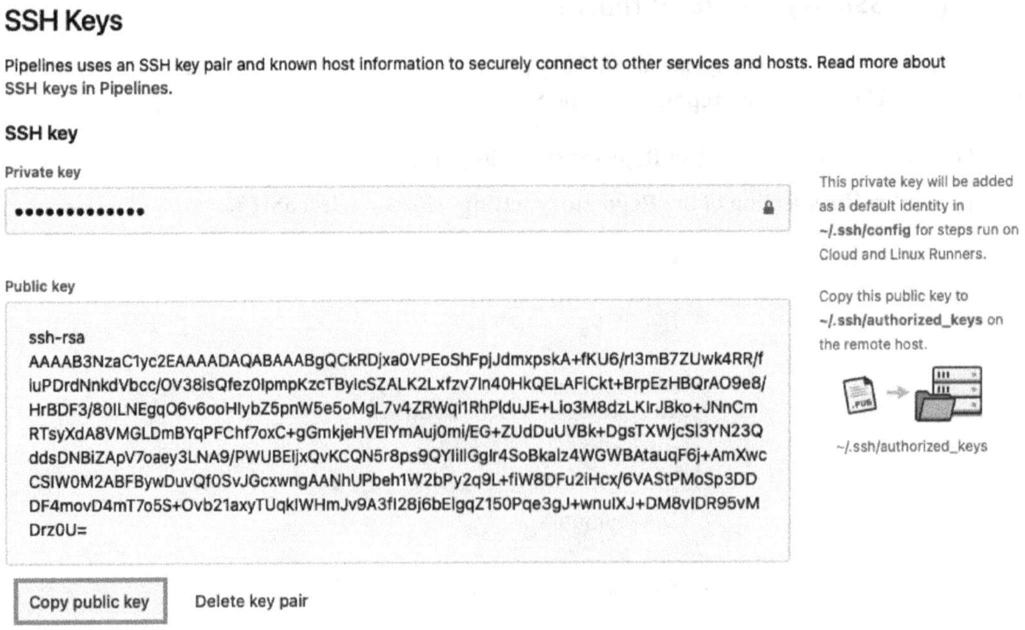

SSH Keys
Pipelines uses an SSH key pair and known host information to securely connect to other services and hosts. Read more about SSH keys in Pipelines.
SSH key
Private key
•••••••••••••
This private key will be added as a default identity in ~/.ssh/config for steps run on Cloud and Linux Runners.
Public key
ssh-rsa
AAAAB3NzaC1yc2EAAAADAQABAAABgQCkRDjxa0VPEoShFpjJdmxpskA+fKU6/rl3mB7ZUwk4RR/f
iuPDrdNnkdVbcc/OV38isQfez0IpmpKzcTBylcSZALK2Lxfzv7In40HkQELAFlCkt+BrpEzHBQrAO9e8/
HrBDF3/80ILNEgqO6v6ooHlybZ5pnW5e5oMgL7v4ZRWqi1RhPlduJE+Lio3M8dzLKlrJBko+JNnCm
RTsyXdA8VMGLDmBYqPFChf7oxC+gGmkjeHVElYmAuj0mi/EG+ZUdDuUVBk+DgsTXWjcSl3YN23Q
ddsDNBiZApV7oaey3LNA9/PWUBEljxQvKCQN5r8ps9QYlillGglr4SoBkalz4WGWBAtauqF6j+AmXwc
CSIW0M2ABFBywDuvQf0SvJGcxwngAANhUPbeh1W2bPy2q9L+fiW8DFu2iHcx/6VAStPMoSp3DD
DF4movD4mT7o5S+Ovb21axyTUqklWHmJv9A3fl28j6bElgqZ150Pqe3gJ+wnulXJ+DM8vlDR95vM
Drz0U=
Copy this public key to ~/.ssh/authorized_keys on the remote host.
.PUB
~/.ssh/authorized_keys
Copy public key
Delete key pair

Figure 8.15 – Copying the public key

5. At this point, you will need to add the public key to your personal Bitbucket settings. To start, select the **Settings** cog and click **Personal Bitbucket settings**:

6. In the **SECURITY** section of the **Personal settings** sidebar, select **SSH keys**:

Figure 8.16 – Selecting SSH keys

7. Select **Add key** on the **SSH keys** page:

Figure 8.17 – Add key

8. In the **Add SSH key** modal, paste the public key and select **Add key**:

Add SSH key

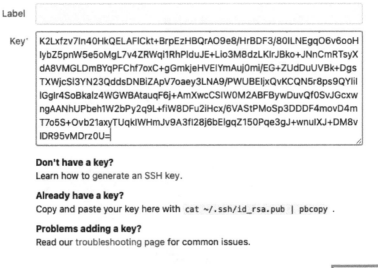

Label

Key* K2Lxfzv7In40HkQELAFICkt+BrpEzHBQrAO9e8/HrBDF3/80ILNEgqO6v6ooH
lybZ5pnW5e5oMgL7v4ZRWqi1RhPlduJE+Lio3M8dzLKlrJBko+JNnCmRTsyX
dA8VMGLDmBYqPFChf7oxC+gGmkjeHVElYmAuj0mi/EG+ZUdDuUVBk+Dgs
TXWjcSI3YN23QddsDNBiZApV7oaey3LNA9/PWUBEljxQvKCQN5r8ps9QYliI
IGglr4SoBkalz4WGWBAtauqF6j+AmXwcCSIW0M2ABFBywDuvQf0SvJGcxw
ngAANhUPbeh1W2bPy2q9L+fiW8DFu2iHcx/6VAStPMoSp3DDDF4movD4m
T7o5S+Ovb21axyTUqkIWHmJv9A3fl28j6bElgqZ150Pqe3gJ+wnuIXJ+DM8v
IDR95vMDrz0U=

Don't have a key?
Learn how to generate an SSH key.

Already have a key?
Copy and paste your key here with `cat ~/.ssh/id_rsa.pub | pbcopy` .

Problems adding a key?
Read our troubleshooting page for common issues.

   Cancel

Figure 8.18 – Pasting and adding the public key

9. The following line needs to be added to the `script` section of your `bitbucket-pipelines.yml` file before the git commands so that you can save and commit changes back to the git repository. This line configures git to use SSH. `BITBUCKET_GIT_SSH_ORIGIN` is a default environment variable:

```
git remote set-url origin ${BITBUCKET_GIT_SSH_ORIGIN}
```

With that, you've configured SSH keys to authenticate when you're pushing a commit back to your git repository.

One final authentication method involves creating an application password and passing that as a secure variable. Let's examine how to do that.

Creating an application password

Application passwords are personal secrets that you can use securely in Bitbucket for automated functions. Let's look at how to create an application password:

1. Select the **Settings** cog and select **Personal Bitbucket settings**.

2. In the **Access Management** section of the **Personal settings** sidebar, select **App passwords**.

3. On the **App passwords** page, select **Create app password**:

Figure 8.19 – Creating an app password

4. On the **Add app password** page, give the app password a name and make sure **Read** and **Write** permissions are selected for the **Repositories** section. Once you've done this, click **Create**:

5. The app password will appear on a modal. Copy the value and make sure it is in a safe location:

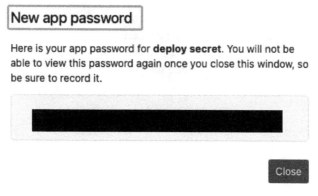

Figure 8.20 – Generated app password

An example of a safe location is a secured variable. Set the app password as a **secured repository variable** or **secured workplace variable**:

Figure 8.21 – Adding the app password as a secured repository variable

6. The following line needs to be added to the `script` section of your `bitbucket-pipelines.yml` file before the git commands so that you can save and commit changes back to the git repository. This line configures the git remote URL to use the included username and app password for authentication:

```
git remote set-url origin https://<your username>:${APP_SECRET}@
bitbucket.org/${BITBUCKET_REPO_OWNER}/${BITBUCKET_REPO_SLUG}
```

With that, we've learned how to use several authentication methods when committing changes back to a git repository. Now, let's look at what needs to be added to do the actual commits.

How to do it...

If you're using the preconfigured HTTP git origin for pushing changes back to the git repository or have set up authentication using one of the preceding methods, the only thing you need to do is define the script steps for performing the actual git commits and pushes. Let's see how that's done:

1. To commit changes, add the `git add`, `git commit`, and `git push` commands to the `script` section of your `bitbucket-pipelines.yml` file. Note that regarding the git commit message, you can add `[skip ci]` to avoid an infinite loop of pipeline executions. An example is shown in the following code snippet:

```
pipelines:
    default:
        - step:
            script:
                - git add <changed files>
                - git commit -m "[skip ci] Updates added via
Bitbucket Pipelines deploy"
                - git push
```

2. If you're using tags, you can add the `git tag` command to create the new tag. The code snippet will resemble the following. In this example, the tag will include the build number since we're referring to the `BITBUCKET_BUILD_NUMBER` predefined variable:

```
pipelines:
    default:
        - step:
            script:
                - git add <changed files>
                - git commit -m "[skip ci] Updates added via
Bitbucket Pipelines deploy"
                - git tag -am "Tag for release ${BITBUCKET_BUILD_
NUMBER}" release-${BITBUCKET_BUILD_NUMBER}
                - git push
```

We have now seen how to use the git repository as a repository for build artifacts. Admittedly, this method is cumbersome when there are dedicated artifact repository tools available such as Artifactory from JFrog and Nexus from Sonatype. In the next recipe, we'll learn how to deploy our build artifacts to those tools for storage and tracking purposes.

Pushing artifacts into artifact repository tools

Artifact repository tools such as Artifactory and Nexus allow for configuration management, a discipline where the build artifacts that emerge from a continuous integration pipeline are stored and tracked against the environments where they are applied.

Elementary builds that result in pushes to Artifactory or Nexus can rely on Bitbucket Pipes, which allows you to easily deploy from the `bitbucket-pipelines.yml` file. More complex deployments involving Maven or npm won't be covered here but we will include references to create those types of deployments.

Getting ready

When using Pipes for either JFrog Artifactory or Sonatype Nexus, there are a few prerequisite steps that should be defined. We will examine which steps are needed by tool.

JFrog CLI prerequisites

The JFrog Setup CLI pipe requires connections to the JFrog Platform servers. Let's look at the steps that are involved in this process:

1. Ensure the runner is installed with the JFrog CLI. At the time of writing, this should be at version `2.17.0` or later. The following command is for macOS and Linux runners, assuming that `curl` is installed:

    ```
    curl -fL "https://getcli.jfrog.io?setup" | sh
    ```

2. If you're using Windows runners, use the following command. Note that this is using PowerShell:

    ```
    powershell "Start-Process -Wait -Verb RunAs powershell
    '-NoProfile iwr https://releases.jfrog.io/artifactory/jfrog-
    cli/v2-jf/[RELEASE]/jfrog-cli-windows-amd64/jf.exe -OutFile
    $env:SYSTEMROOT\system32\jf.exe'" ; jf setup
    ```

3. Use the JFrog CLI to connect to your JFrog Platform servers. To set up the details of your JFrog Platform instances, run the following command:

    ```
    jf c add
    ```

    Executing this command prompts you to create a server ID and associate it with the URL for your JFrog platform. It is recommended that you set up a repository variable that maps to the environment variable that's used by JFrog – that is, `JFROG_CLI_SERVER_ID`.

4. At this point, it may also be convenient to set up Bitbucket secure variables that mirror JFrog environment variables that have names starting with `JF_ENV_` as the key and the server token as the value. You can derive the server token using the following JFrog CLI command:

```
jf c export <server ID from previous instruction>
```

With that, we've covered the preliminary setup for the JFrog CLI. Now, let's examine what's needed for Nexus.

Sonatype Nexus prerequisites

The pipe for Sonatype Nexus Publisher requires a few environment variables to be defined. Some variables are mandatory. The variables to be defined as follows:

- `FILENAME` (mandatory): This is the path to the file to publish.
- `ATTRIBUTES` (mandatory): Attributes needed by Nexus Publisher. Component attributes are denoted with `-C`. Asset attributes are denoted with `-A`.
- `USERNAME` (mandatory): Nexus username.
- `PASSWORD` (mandatory): Nexus password.
- `SERVER_URL` (Mandatory): Nexus server URL.
- `REPOSITORY`: Repository name in Nexus. The default is `maven-releases`.
- `FORMAT`: Artifact format. The default is `maven2`.

Now that our prerequisites are out of the way, let's look at pushing our artifacts using pipes.

How to do it...

In the *Connecting to Bitbucket Pipes* recipe in *Chapter 6*, we learned about pipes. Pipes serve as integration points for third-party tools in a Bitbucket pipeline.

At this point, we are ready to connect to either JFrog or Sonatype Nexus through pipes and set up deployment steps. Let's examine how to do that for each tool.

Using the JFrog Setup CLI pipe

You can use the JFrog Setup CLI pipe to connect to any JFrog tool on the JFrog Platform, including Artifactory and XRay. Let's look at the steps involved:

1. Add the following lines to the `script` section of your `bitbucket-pipelines.yml` file:

```
script:
    - pipe: jfrog/jfrog-setup-cli:2.0.0
    - source ./jfrog-setup-cli.sh
```

2. Once the setup lines have been added, you can use any JFrog CLI commands to make changes. The following code snippet is an example of a build that runs several commands to Artifactory using JFrog CLI commands:

```
script:
    - pipe: jfrog/jfrog-setup-cli:2.0.0
    - source ./jfrog-setup-cli.sh
    # Upload artifacts to Artifactory
    - jf rt u file artifacts/
    # Collect environment variables
    - jf rt bce
    # Publish build info
    - jf rt bp
```

With that, we've learned how to connect and communicate with JFrog tools using pipes and JFrog CLI commands. Now, let's look at the process for Sonatype Nexus Manager.

Using the Sonatype Nexus Publisher pipe

Let's learn how to use a pipe to connect to Sonatype Nexus Publisher:

1. Add the following code snippet to the `script` section of your `bitbucket-pipelines.yml` file. You can fill out the variables in-line or use Bitbucket variables:

```
- pipe: sonatype/nexus-repository-publish:0.0.1
  variables:
    FILENAME: '<string>'
    ATTRIBUTES: '<string>'
    USERNAME: '<string>'
    PASSWORD: '<string>'
    SERVER_URL: '<string>'
    # REPOSITORY: '<string>' # Optional.
    # FORMAT: '<string>' # Optional.
```

2. Here's an example:

```
    - step:
        # set NEXUS_USERNAME and NEXUS_PASSWORD as environment
variables
        name: Deploy to Nexus Repository Manager
        deployment: test    # set to test, staging or production
        # trigger: manual   # uncomment to have a manual step
        script:
          - pipe: sonatype/nexus-repository-publish:0.0.1
            variables:
              FILENAME: 'target/myapp-1.0-SNAPSHOT.jar'
```

```
            ATTRIBUTES: '-CgroupId=com.example
  -CartifactId=myapp -Cversion=1.0 -Aextension=jar'
            USERNAME: '$NEXUS_USERNAME'
            PASSWORD: '$NEXUS_PASSWORD'
            SERVER_URL: 'https://nexus.example.com/'
```

At this point, we've seen how easy it is to use pipes for easy deployment. Other examples in subsequent recipes in this chapter will also show the ease of pipes.

See also

The following are links to other reference materials if you need to learn more about how to connect to JFrog or Sonatype Nexus:

- JFrog CLI documentation: `https://docs.jfrog-applications.jfrog.io/jfrog-applications/jfrog-cli`

- An example of using Bitbucket Pipelines for a Maven deployment using **GNU Privacy Guard (GPG)** and **Open Source Software Repository Hosting (OSSRH)**: `https://bitbucket.org/simpligility/ossrh-pipeline-demo/src/master/`

Next, we'll look at several examples of taking build artifacts and installing them on target environments. We'll begin by looking at various upload processes that can be used.

Deploying artifacts to Bitbucket Downloads

You can use Bitbucket itself as a staging area for build artifacts by configuring the Bitbucket Downloads pipe. This allows you to push build artifacts to Bitbucket Downloads. We'll examine this process in more detail in this recipe.

Getting ready

Use of the `bitbucket-upload-file` pipe requires authentication either by username and app password or by access token. We looked at creating app passwords in the *Getting ready* section of the *Pushing artifacts into the Bitbucket repository* recipe.

Access tokens are available at the repository level for all plans of Bitbucket, and they're available at the project and workspace levels for the Premium plan of Bitbucket. The tokens are only scoped to the repository, project, or workspace for which they were created. These are for single-use functions and are revoked if replacement is needed. Let's learn how to create a repository access token:

1. At the repository level, select **Repository settings** in the sidebar.
2. In the **Repository settings** sidebar, select **Access tokens** in the **SECURITY** section:

Figure 8.22 – Selecting Access tokens

3. On the **Access tokens** page, select **Create Repository Access Token**:

4. In the modal, give the token a name and specify permissions. Bitbucket Pipelines requires **Read** and **Write** permissions on **Repositories**. Click **Create** when you're finished:

5. The next modal will contain the value of the token and useful applications for it. Save the token value by copying it and pasting it in a safe location. This will be the only opportunity you will have to view the token value.

6. You can place your access token as a secure repository variable. Select **Repository variables** in the **Repository settings** sidebar, fill in the key name for the token, and paste the previously copied token value into the **Value** section. Once you've done this, check the **Secured** checkbox and click **Add**:

Repository variables

Environment variables added on the repository level can be accessed by any users with push permissions in the repository. To access a variable, put the $ symbol in front of its name. For example, access AWS_SECRET by using $AWS_SECRET.
Learn more about repository variables.

Repository variables override variables added on the workspace level. View workspace variables

If you want the variable to be stored unencrypted and shown in plain text in the logs, unsecure it by unchecking the checkbox.

You haven't added any variables yet.

Figure 8.23 – Repository variables

Now that we have established our authentication means, either by username/app password or access token, let's set up our pipe to Bitbucket Downloads.

How to do it...

The `bitbucket-upload-file` pipe is the primary means of deploying files into the Bitbucket Downloads area. Let's learn how to set up that pipe:

1. If you want to authenticate using a username and app password, use the following code snippet:

    ```
    script:
      - pipe: atlassian/bitbucket-upload-file:0.7.1
        variables:
    ```

```
      BITBUCKET_USERNAME: $BITBUCKET_USERNAME
      BITBUCKET_APP_PASSWORD: $BITBUCKET_APP_PASSWORD
      FILENAME: 'package.json'
```

2. To use the access token instead, replace `BITBUCKET_USERNAME` and `BITBUCKET_APP_PASSWORD` with `BITBUCKET_ACCESS_TOKEN`. This is illustrated in the following code snippet:

```
script:
  - pipe: atlassian/bitbucket-upload-file:0.7.1
    variables:
      BITBUCKET_ACCESS_TOKEN: $BITBUCKET_ACCESS_TOKEN
      FILENAME: 'package.json'
```

3. The `FILENAME` pipe variable can specify multiple files by invoking wildcards. Note that the limit is 10 files. The following code shows an example of uploading all `.txt` files:

```
script:
  - pipe: atlassian/bitbucket-upload-file:0.7.1
    variables:
      BITBUCKET_USERNAME: $BITBUCKET_USERNAME
      BITBUCKET_APP_PASSWORD: $BITBUCKET_APP_PASSWORD
      FILENAME: '*.txt'
```

4. Other optional pipe variables allow you to specify another account and repository where the file will be uploaded. `ACCOUNT` and `REPOSITORY` are illustrated in the following code snippet:

```
script:
  - pipe: atlassian/bitbucket-upload-file:0.7.1
    variables:
      BITBUCKET_USERNAME: $BITBUCKET_USERNAME
      BITBUCKET_APP_PASSWORD: $BITBUCKET_APP_PASSWORD
      FILENAME: 'package.json'
      ACCOUNT: $PROJECT_ACCOUNT
      REPOSITORY: $ALTERNATE_REPO
```

With that, we've learned how to push files to the Bitbucket Downloads area for retrieval. Now, let's learn how to send build artifacts to target servers.

Deploying artifacts using SCP

SCP is a means of transferring files between two host computers. This protocol uses SSH as a foundation to securely move files from one computer to another.

Let's learn how to transfer a build artifact from Bitbucket Cloud to the remote host using SCP.

Getting ready

Because the foundation of SCP is SSH, we need to prepare an SSH key and other configurations related to SSH on both Bitbucket and the remote host. Let's take a look at the necessary steps:

1. We defined a repository SSH key in the *Getting ready* section of the *Pushing artifacts into the Bitbucket repository* recipe. We can use this key for transfer to the remote host by copying the public key and placing it in the `~/.ssh/authorized_keys` file. If you have SSH access to the remote host, run the following command from the machine where you generated the key pair. This mandates that the user performing the operation will be you:

```
ssh-copy-id -i <public key file to copy> user@host
```

2. We also need to update the known hosts on Bitbucket. For the repository, this is located on the same screen we used to create the SSH key. From **Repository Settings**, select **SSH Keys**. On the **SSH Keys** screen, type in the IP address for the remote host and click **Fetch**.

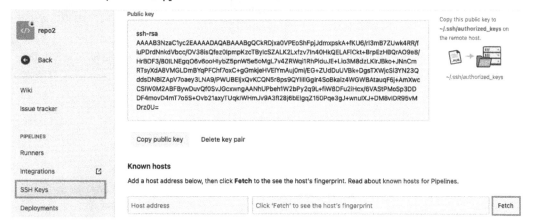

Figure 8.24 – Adding a known remote host

3. Configure your remote host so that it allows SCP/SSH access on your desired port (`22` is the default) and allows access using SSH keys. This step is left for you as an exercise because of the variety of systems and configurations that are available.

Once we have the necessary configurations, it's time to configure the `bitbucket-pipelines.yml` file.

How to do it...

Now that we've configured the SSH keys both on Bitbucket Cloud and the remote host, we can set up Bitbucket Pipelines to make the file transfer process a deployment:

1. Add the following code snippet to the `script` section of your `bitbucket-pipelines.yml` file. The necessary pipe variables include the user's name for the remote host, the remote host name, the path on the remote host to deploy files to, and the local path where the build artifacts are located:

```
- pipe: atlassian/scp-deploy:1.5.0
  variables:
    USER: '<string>'
    SERVER: '<string>'
    REMOTE_PATH: '<string>'
    LOCAL_PATH: '<string>'
    # SSH_KEY: '<string>' # Optional.
    # EXTRA_ARGS: '<string>' # Optional.
    # DEBUG: '<boolean>' # Optional.
```

2. You can also add options using the `EXTRA_ARGS` pipe variable. An example is shown in the following code snippet:

```
script:
  - pipe: atlassian/scp-deploy:1.5.0
    variables:
      USER: 'ec2-user'
      SERVER: '127.0.0.1'
      REMOTE_PATH: '/var/www/build/'
      LOCAL_PATH: 'build/'
      DEBUG: 'false'
      EXTRA_ARGS: ["-P", "8022"]
```

3. The `SSH_KEY` pipe variable allows you to define an alternate SSH key. This should be a base64-encoded private key, saved as a secured Bitbucket variable. The following code snippet shows the use of an alternate SSH key:

```
script:
  - pipe: atlassian/scp-deploy:1.5.0
    variables:
      USER: 'ec2-user'
      SERVER: '127.0.0.1'
      REMOTE_PATH: '/var/www/build/'
      LOCAL_PATH: 'build'
      SSH_KEY: $MY_SSH_KEY
      DEBUG: 'true'
      EXTRA_ARGS: ['-o', 'ServerAliveInterval=10']
```

With that, we've deployed to target environments using Bitbucket Pipelines. So far, the target environments are physical servers or virtual machines. Next, we'll learn how to deploy to public cloud environments.

Deploying artifacts into AWS S3 buckets

In this recipe, we're going to look at deploying to AWS. Bitbucket Pipelines has a variety of pipes that can deploy to specific AWS component services, depending on the type of build artifact.

Let's look at what's involved in deploying our build artifact into an AWS S3 bucket. This is a typical use case of deployment using Bitbucket Pipelines.

Getting ready

When setting up the pipe to deploy to an AWS S3 bucket, the only thing you need to do is set up the following variables:

- `AWS_ACCESS_KEY_ID`: Your AWS access key.

- `AWS_SECRET_ACCESS_KEY`: Your AWS secret access key. This should be saved as a secured variable.

- `AWS_DEFAULT_REGION`: The default AWS region of your resource.

Now that these have been defined, let's look at setting up the pipe.

How to do it...

At this point, we need to add our pipe configuration to the `script` section of our `bitbucket-pipelines.yml` file. Let's take a closer look:

1. Add the following code snippet to the `script` section of the `bitbucket-pipelines.yml` file. This will contain the variables that you set in the *Getting ready* section incorporated into the `script` section. Here, `S3_BUCKET` defines the destination bucket and `LOCAL_PATH` defines the location of the build artifact:

```
script:
  - pipe: atlassian/aws-s3-deploy:1.6.0
    variables:
      AWS_ACCESS_KEY_ID: $AWS_ACCESS_KEY_ID
      AWS_SECRET_ACCESS_KEY: $AWS_SECRET_ACCESS_KEY
      AWS_DEFAULT_REGION: 'us-east-1'
      S3_BUCKET: 'my-bucket-name'
      LOCAL_PATH: 'build'
```

2. You can also define a folder inside the bucket by appending the path to the bucket defined in S3_BUCKET. This is illustrated in the following code snippet:

```
script:
  - pipe: atlassian/aws-s3-deploy:1.6.0
    variables:
      AWS_ACCESS_KEY_ID: $AWS_ACCESS_KEY_ID
      AWS_SECRET_ACCESS_KEY: $AWS_SECRET_ACCESS_KEY
      AWS_DEFAULT_REGION: 'us-east-1'
      S3_BUCKET: 'my-bucket-name/logs'
      LOCAL_PATH: '$(pwd)'
```

With that, you've seen how easy it is to connect Bitbucket Pipelines so that you can deploy build artifacts to AWS. Let's examine doing the same thing for Google Cloud.

Deploying artifacts to Google Cloud

Bitbucket Pipelines can make deployments to Google services through the use of pipes that integrate with these services. The following is the current list of pipes that connect with Google services:

- Firebase deploy
- Google App Engine Deploy
- Google Cloud Storage Deploy
- **Google Artifactory Registration (GAR)** push image
- Google Kubernetes Engine kubectl run

Let's take a closer look at deploying build artifacts using the Google Cloud Storage Deploy pipe.

How to do it...

We can deploy our artifacts to Google Cloud Storage by performing the following steps:

1. In the script portion of the step area where you intend to deploy to Google Cloud Storage, copy and paste the pipe definition. The definition of the Google Cloud Storage Deploy pipe is shown in the following code snippet:

```
- pipe: atlassian/google-cloud-storage-deploy:2.0.0
  variables:
    KEY_FILE: '<string>'
    BUCKET: '<string>'
    SOURCE: '<string>'
    # GOOGLE_OIDC_CONFIG_FILE: "<string>" # Optional by default.
Required for OpenID Connect (OIDC) authentication.
    # PROJECT: "<string>" # Optional by default. Required with
```

```
GOOGLE_OIDC_CONFIG_FILE.
    # CACHE_CONTROL: '<string>' # Optional. options include
no-cache,no-store,max-age=<seconds>, s-maxage=<seconds>,
no-transform, public, private
    # CONTENT_DISPOSITION: '<string>' # Optional.
    # CONTENT_ENCODING: '<string>' # Optional.
    # CONTENT_LANGUAGE: '<string>' # Optional.
    # CONTENT_TYPE: '<string>' # Optional.
    # ACL: '<string>' # Optional.  Options include project-
private, private, public-read, public-read-write, authenticated-
read, bucket-owner-read, bucket-owner-full control
    # STORAGE_CLASS: '<string>' # Optional.  Options include
multi-regional, regional, nearline, coldline
    # DEBUG: '<boolean>' # Optional.
```

2. Add the pipe to the `script` section of your `bitbucket-pipelines.yml` file. An example with only the mandatory pipe variables provided is shown in the following code snippet:

```
script:
  - pipe: atlassian/google-cloud-storage-deploy:2.0.0
    variables:
      KEY_FILE: $KEY_FILE
      BUCKET: 'my-bucket'
      SOURCE: 'myApp.jar'
```

3. If needed, continue defining the deployment operation by adding other variables. The following code snippet shows a pipe with more variables filled out:

```
script:
  - pipe: atlassian/google-cloud-storage-deploy:2.0.0
    variables:
      KEY_FILE: $KEY_FILE
      BUCKET: 'my-bucket'
      SOURCE: 'myAppFile.jar'
      CACHE_CONTROL: 'max-age=60'
      CONTENT_DISPOSITION: 'attachment'
      ACL: 'public-read'
      STORAGE_CLASS: 'nearline'
```

We have just seen an example of integrating a deployment to Google Cloud resources using pipes. Now, let's look at an example of integration with Microsoft Azure.

Deploying artifacts to Microsoft Azure

Bitbucket Pipelines can perform deployments to Azure services through the use of pipes that integrate with these services. The following is the current list of pipes that connect with Microsoft Azure:

- Azure CLI
- Azure Container Apps Deploy
- Azure ACR push image
- Azure Functions Deploy
- Azure Kubernetes Service Deploy
- Azure Kubernetes Service Helm Deploy
- Azure Storage Deploy
- Azure Web Apps Containers Deploy
- Azure Web Apps Deploy

Let's see what's needed to use the Azure Functions Deploy pipe as an example of deployment to Microsoft Azure.

Getting ready

Before we can use the Azure Functions Deploy pipe to deploy to Microsoft Azure, we need to establish our Azure credentials. Follow these steps to do so:

1. After installing the Azure CLI on your local machine or using the Azure Cloud Shell, create an Azure secure principal by typing the following command:

    ```
    az ad sp create-for-rbac --name <name of your service principal>
    ```

2. The preceding command will return the following output in JSON format:

    ```
    {
        "appId": "myAppId",
        "displayName": "myServicePrincipalName",
        "password": "myServicePrincipalPassword",
        "tenant": "myTentantId"
    }
    ```

3. Save the output as repository variables. For instance, you can define the following output as variables in this way:

- `appId`: `AZURE_ID`

- `password`: `AZURE_PASSWORD` (secured)

- `tenant`: `AZURE_TENANT`

Now that we have our secure principal, we can connect to Azure using Bitbucket Pipelines. Let's learn how to do that.

How to do it...

Azure Functions Deploy takes serverless logic implementations written on your favorite Microsoft development tools, such as Visual Studio, and packages them to be executed on-demand in Azure. Part of the development process can include deployment through Bitbucket Pipelines. Let's learn how to deploy functions from the `bitbucket-pipelines.yml` file:

1. Add the following code snippet to the `script` section of `bitbucket-pipelines.yml`. The required parameters include the attributes for the secure principal, the function name, as found in Azure, and the name of the ZIP file that contains the function to be deployed on Azure:

```
script:
  - pipe: atlassian/azure-functions-deploy:2.0.0
    variables:
      AZURE_APP_ID: $AZURE_APP_ID
      AZURE_PASSWORD: $AZURE_PASSWORD
      AZURE_TENANT_ID: $AZURE_TENANT_ID
      FUNCTION_APP_NAME: '<string>'
      ZIP_FILE: '<string>'
      # DEBUG: '<boolean>' # Optional
```

2. Continue adding the necessary variables to the pipe definition. A completed example can be seen here:

```
script:
  - pipe: atlassian/azure-functions-deploy:2.0.0
    variables:
      AZURE_APP_ID: $AZURE_APP_ID
      AZURE_PASSWORD: $AZURE_PASSWORD
      AZURE_TENANT_ID: $AZURE_TENANT_ID
      FUNCTION_APP_NAME: 'my-function'
      ZIP_FILE: 'application.zip'
```

With that, we've deployed a serverless function from our local system to Azure.

One aspect of deployment is **Infrastructure as Code (IaC)** or the use of text-based configurations to dynamically create needed resources. A popular tool for doing this is Ansible. In the next recipe, we'll learn how to use Bitbucket Pipelines to execute Ansible playbooks and deploy resources.

Using Ansible in the deployment stage

Ansible is a standard tool for performing IaC. With Ansible, you can configure physical or virtual servers and perform configuration tasks such as installing or upgrading software, setting the necessary parameters, and starting application services.

Ansible is available in its original CLI as well as an integrated graphical user interface for dedicated Ansible application servers called **Ansible Tower**. We will look at automating deployment to both interfaces.

Getting ready

To execute Ansible, the following programs must be installed on the runners you plan to use to execute the Bitbucket pipeline:

- `python`
- `pip`

Once they're installed, we can add Ansible commands to our Bitbucket pipeline.

How to do it...

The original Ansible application is built on Python and accepts two files as input. Both of these files are text files in YAML format:

- `playbook`: This file includes the commands that Ansible runs
- `Inventory`: This file details the machines and their environments that Ansible applies the playbook against

Let's learn how to incorporate this in our `bitbucket-pipelines.yml` file:

1. Add the following line to the `script` section of the `bitbucket-pipelines.yml` file. This will install Ansible:

    ```
    - pip install ansible==2.17
    ```

2. Once installed, move inside the directory where the playbooks for Ansible reside:

    ```
    cd deployment
    ```

3. Add the command to run Ansible. You can use the `-i` flag to denote the inventory file:

```
- ansible-playbook -i inventory playbook.yaml
```

4. The script should look as follows:

```
script:
    - pip install ansible==2.17
    - cd deployment
    - ansible-playbook -i inventory playbook.yaml
```

In this recipe, we looked at how to deploy a configuration from Bitbucket Pipelines using the community version of Ansible. The full version from Red Hat features dedicated infrastructure for running the Ansible application that's controlled by a GUI called Ansible Tower. Next, we'll learn how to deploy to Ansible Tower using Bitbucket Pipelines.

There's more...

Although the primary means of controlling Ansible Tower jobs is through a GUI, there is a CLI called `tower-cli` that allows you to script Ansible Tower jobs. Let's look at `tower-cli` in action:

1. Install `tower-cli` in your build environment. You may need to specify an image that has Python, including `pip`. Here, `pip` allows you to install `tower-cli`:

```
image: python:2.7
  pipelines:
   default:
    - step:
         script: # Modify the commands below to build your
repository.
               - pip install ansible-tower-cli
```

2. Set up the required environment variables. For `tower-cli`, these are as follows:

- host: Tower host

- username: Tower username

- password: Tower user password (saved as a secure Bitbucket variable)

- ID: ID of the Tower job template to launch

3. Add the following lines to configure the environment variables and run the Ansible job:

```
- hostval=$(tower-cli config host $host)
- userval=$(tower-cli config username $username)
- passwordval=$(tower-cli config password $password)
- tower-cli config verify_ssl false
- tower-cli job launch --job-template $ID --monitor
```

With that, we've triggered an Ansible job located in Ansible Tower from Bitbucket Pipelines.

See also

The following documentation for Ansible provides guidance on creating correct Ansible jobs:

- https://access.redhat.com/documentation/en-us/red_hat_ansible_automation_platform/2.4

- https://docs.ansible.com/ansible/latest/index.html

Using Terraform in the deployment stage

Terraform is another tool that provides IaC capabilities. Its popularity comes from the fact that it is flexible for outlining instance creation of cloud resources and then implementing those resources to specific cloud platforms using providers that specify the implementation details.

Let's learn how to deploy to Terraform from Bitbucket Pipelines.

Getting ready

Terraform describes the configuration it will perform in three files, all of which need to be in your Bitbucket repository:

- `main.tf`

- `variables.tf`

- `provider.tf`

In addition, any credentials needed by Terraform for configuring the backend platforms, such as AWS or Google Cloud, should be stored as Bitbucket variables, and you should allocate them as secure if necessary.

With these in place, let's set up our deployment.

How to do it...

There are several steps we must take to deploy using Terraform, encapsulated in three commands. Let's look at the Terraform commands we need to use:

1. For the step that defines deployment, use the Terraform Docker image. We'll cover using Docker images in Bitbucket Pipeline steps in more detail in *Chapter 9*:

   ```
   - step:
       image: hashicorp/terraform:full
   ```

2. In the `script` section of the `bitbucket-pipelines.yml` file, enter the following command to initialize Terraform:

   ```
   - terraform init
   ```

3. Add the following line to perform validation. This may not be required:

   ```
   - terraform validate
   ```

4. Add the following line to create the plan in Terraform. The output can be saved using the `-out` flag:

   ```
   - terraform plan -out=plantf
   ```

5. Add the following line to apply the plan and run Terraform:

   ```
   - terraform apply plantf
   ```

6. The complete `step` should now look as follows:

   ```
   - step:
       image: hashicorp/terraform:full
       script:
         - terraform init
         - terraform validate
         - terraform plan -out=plantf
         - terraform apply plantf
   ```

With that, we performed a deployment in Bitbucket Pipelines where we invoked Terraform to create our instances.

See also

The following resource is beneficial for understanding Terraform – `https://developer.hashicorp.com/terraform/docs`.

9

Leveraging Docker and Kubernetes for Advanced Configurations

In the previous chapter, we looked at implementing continuous deployment through Bitbucket Pipelines and worked with various platforms using various technologies. However, we reserved discussions of deploying with Docker and Kubernetes until now. Bitbucket Pipelines can leverage containers for its build environment, as a build package, and even as runners to run pipeline executions. In each case mentioned, you can use a public image or create and use a private image.

In this chapter, we will examine using Docker container technology and Kubernetes. We'll cover the following recipes in this chapter while implementing Bitbucket Pipelines:

- Using a Docker image as a build environment
- Using containerized services in Bitbucket Pipelines
- Using Docker commands in Bitbucket Pipelines
- Deploying a Docker image to Kubernetes using Bitbucket Pipelines
- Setting up Docker-based runners in Linux

Let's begin our exploration of Docker and Kubernetes in Bitbucket Pipelines.

Technical requirements

Before we begin our exploration, we should identify what's needed to work with Docker and Kubernetes in our local development environment.

To work with **Docker images**, you need to make sure Docker applications are installed on your runner machines for executing any Docker commands.

For development machines that are used to create build environments, a good option is Docker Desktop, an application that provides all the required Docker tools for building, packaging, running, and deploying applications as containers. It is available for Mac, Windows, and Linux machines. More information can be found at `https://docs.docker.com/get-docker/`.

Runners only require Docker applications that can build and run containerized applications. For that, Docker Engine is a good application to install and configure on your runner. It is available for many common Linux distributions, including Ubuntu, Debian, and Red Hat. More information can be found at `https://docs.docker.com/engine/`.

For working with Kubernetes clusters, `kubectl` is the preferred tool. The binary for `kubectl` can be downloaded for Linux, Mac, or Windows. Package managers such as yum for Red Hat Linux, apt for Debian Linux, homebrew for Mac, and chocolatey for Windows can also download and install `kubectl`. More information can be found at `https://kubernetes.io/docs/home/`.

The sample code for this chapter can be found in the `Chapter9` folder of this book's GitHub repository `https://github.com/PacktPublishing/Atlassian-DevOps-Toolchain-Cookbook/tree/main/Chapter9`

Introducing containers and Bitbucket Pipelines

One of the most recent advances in technology that furthered the DevOps movement was the introduction of **container** technology. As mentioned in *Chapter 1*, instead of setting up complete environments as physical or **virtual machines (VMs)**, an application and its required libraries would reside in a self-contained entity called a container. This container interacts with outside resources through a managing application. At the time of writing, the most popular application for managing containers is Docker Engine from Docker Inc.

Containers have allowed for application portability at an unprecedented level. Developers can create an application, package it as a container, and run tests on the application in a test environment managing that container. Deployment to production would use the same container image, but in an environment with possibly more resources, depending on the target, allowing multiple instances of the application container for load sharing or high availability.

Bitbucket Pipelines can work with containers. So, let's consider some of the uses of containers in Bitbucket Pipelines.

By default, Bitbucket Pipelines uses Docker images as build. You can define which Docker image to use for your build.

You can also use Bitbucket Pipelines to create a Docker container image and update the appropriate Docker container repository.

Bitbucket Pipelines can use runners that can be created from Docker images to execute builds. This may allow for dynamic allocation of runners to perform a build by creating as many Docker containers as needed and destroying those containers when complete. The only limitation to this is the available resources in your environment.

Another application of container technology comes in the form of Kubernetes. Kubernetes was initially developed by Google to abstract applications stored in containers as services and provide an environment for establishing and maintaining clusters of containerized services.

Finally, Bitbucket Pipelines can build an application into a Docker container image. This image can be deployed into a Kubernetes cluster as part of the pipeline script.

Now that we understand how Docker containers work with Bitbucket Pipelines, let's examine how we can make that happen.

Using a Docker image as a build environment

Bitbucket Pipelines uses a Docker image as a platform to execute the commands found in `bitbucket-pipelines.yml`. This image is normally a default provided by Atlassian but can be replaced with a custom image.

Let's examine how Bitbucket Pipelines uses these Docker images.

Getting ready

In Bitbucket Pipelines, the runner executes the commands specified on `bitbucket-pipelines.yml` in a build environment. This build environment always uses a Docker container.

If a Docker image isn't specified, Bitbucket Pipelines will select a default Docker image for the container.

Default images used by Bitbucket Pipelines are stored by Atlassian in **Docker Hub** at `https://hub.docker.com/r/atlassian/default-image/`.

Version numbers for the default image can be specified. If no version number is specified, the version specified with the **latest** tag will be used.

The following table provides a synopsis of the default image versions:

Version	Tags	Contents
1.x (deprecated)	`latest`	Platform: `ubuntu 14.04` Packages available out of the box: `wget xvfb curl git: 1.9.1 java: 1.8u66 maven: 3.0.5 node: 4.2.1 npm: 2.14.7 nvm: 0.29.0 python: 2.7.6 gcc: 4.8.4`
2.x (deprecated)		Platform: `ubuntu 16.04` Packages available out of the box: `wget xvfb curl ssh git: 2.7.4 mercurial: 3.7.3 java: Open-JDK 1.8u151 maven: 3.3.9 node: 8.9.4 npm: 5.6.0 nvm: 0.33.8 python: 2.7.12 gcc: 5.4.0 ant: 1.9.6`

Version	Tags	Contents
3.x (deprecated)		Platform: `ubuntu 20.04 (LTS)` Packages available out of the box: `wget xvfb curl ssh zip jq tar parallel git: 2.39.1 node: 14.17.5 npm: 6.14.14 nvm: 0.38.0 python: 3.8.10 gcc: 9.4.0 ant: 1.10.7`
4.x (recommended)		Platform: `ubuntu 22.04 (LTS)` Packages available out of the box: `wget xvfb curl ssh zip jq tar parallel git: 2.39.1 node: 18.16.1 npm: 9.5.1 nvm: 0.39.2 python: 3.10.6 gcc: 11.3.0 ant: 1.10.12`

Table 9.1 – Default Atlassian build environment Docker images

> **Important note**
>
> An image tagged as `latest` is using an older image with other images that are created more recently. This allows for backward compatibility with older Bitbucket Pipelines builds.

To specify the desired version, add the following line to `bitbucket-pipelines.yml`:

```
image: atlassian/default-image:<version number>
```

Here, `<version number>` represents the desired version or tag (such as `latest`).

We can specify any Docker image, from a public or private repository, to create our build environment. Let's examine the methods for doing so in our next section.

How to do it...

Bitbucket Pipelines can use any Docker image from public or private repositories. The information that's required differs based on whether the repository is public or private.

Let's learn how to use a Docker image from a public registry.

Using a public image

A public repository hosts Docker images available for use by anyone. This repository can reside on Docker Hub, another repository, or even a self-published repository, so long as it can be accessed on the internet.

Let's look at using a public Docker image for your build environment:

1. Specify the image by name in the `bitbucket-pipelines.yml` file. If a tag isn't included, the `latest` tag is implied:

```
image: postgres
```

2. If an account is specified, it should be included as part of the name:

```
image: bitnami/postgresql
```

3. Specific versions can be included after the image's name with a colon:

```
image: bitnami/postgresql:16.2.0
```

4. If you're using a public image not hosted on Docker Hub, include the repository's URL in the image specification:

```
image: docker.publicimage.com/bitnami/postgresql:16.2.0
```

With that, you've learned how to specify public Docker images for build environments. Now, let's look at using private images.

Using a private image

Private Docker repositories are often used by companies and other organizations to store custom Docker images containing that organization's intellectual property. These repositories are often secured through authentication policies.

Let's look at configuring build environments by using private Docker images:

1. You can add secured variables to store your credentials and refer to the variables in the `bitbucket-pipelines.yml` file. Variables and secrets were discussed in *Chapter 6*. The following code snippet shows an example of a private Docker Hub repository:

```
image:
    name: my-company-account/bitnami/postgresql:16.2.0
    username: $DOCKER_HUB_USERNAME
    password: $DOCKER_HUB_PASSWORD
    email: $DOCKER_HUB_EMAIL
```

2. If your private Docker repository uses AWS **EC2 Container Registry** (**ECR**), you can set up the access key and secret key as variables. They will be identified in a separate `aws` section:

```
image:
    name:
    <aws_account_id>.dkr.ecr.<region>.amazonaws.com/bitnami/
    postgresql:16.2.0
```

```
aws:
    access-key: $AWS_ACCESS_KEY
    secret-key: $AWS_SECRET_KEY
```

3. Another method of passing AWS credentials involves setting up an IAM role in AWS and setting up Bitbucket Pipelines as a web identity provider. This allows Bitbucket Pipelines to connect to AWS ECR using Open ID Connect. Detailed instructions for this are located at `https://support.atlassian.com/bitbucket-cloud/docs/use-aws-ecr-images-in-pipelines-with-openid-connect/`. The following code snippet must then be placed in `bitbucket-pipelines.yml`:

```
image:
    name: <aws_account_id>.dkr.ecr.<region>.amazonaws.com/
bitnami/postgresql:16.2.0
    aws:
        oidc-role: arn:aws:iam::<aws_account_id>:role/<your_
role_name>
```

4. If your private Docker repository is located in **Google Container Registry** (**GCR**), you must create a service account in the GCP admin console that grants *Viewer* access to GCR for Bitbucket Pipelines. This will create a private key in JSON format. Download the key and save it in Bitbucket Pipelines as a secured variable. You can then access the image using the following code snippet:

```
image:
    name: <region>.gcr.io/<project>/image
    username: _json_key
    password: '$GCR_JSON_KEY'
```

For any other private Docker repository, provide the registry URL and include the credentials as secured variables. This is shown in the following code snippet:

```
image:
    name: docker.<company name>.com/<account-name>/bitnami/
postgresql:16.2.0
    username: $USERNAME
    password: $PASSWORD
    email: $EMAIL
```

With that, you've seen a variety of places where Bitbucket Pipelines can retrieve Docker images for use as build environments.

Next, we'll look at defining and using containerized services while running our Bitbucket pipeline.

Using containerized services in Bitbucket Pipelines

You can run multiple services in a Bitbucket pipeline by defining containers to use. Once the pipeline runs, these services are scheduled to run in the step they are invoked. Services that can be invoked in this manner include databases, code analytics, and web services.

In this recipe, we'll look at defining and using containerized services.

Getting ready

There are a few things to understand about the limitations of using services implemented in containers during pipeline executions. Let's take a close look at them now.

There are a limited number of resources available for these containerized services. Any given step in the pipeline can work with a maximum of five services. If you need to run with a larger number of services, you can define a **Docker-in-Docker** configuration that allows you to execute additional services through `docker run` or `docker-compose`.

Each of these services will run without waiting for service startup. While these services are running, you cannot access the services or their logs through REST API calls, although logs should be available through the Bitbucket Pipelines **user interface (UI)**. Also, while running, TCP and UDP port `29418` will be reserved and cannot be used for external actions.

The most involved limits for services involve memory. Each step can be defined as either a regular step with a 4,096 MB memory limit or as a large build step (defined by adding the `size: 2x` statement in the step definition), which increases the memory limit to 8,192 MB.

The memory in a step is divided into one build container and the number of service containers as defined by the step. A build container requires a minimum of 1,024 MB. This amount of memory is needed to handle the build process and any overhead required by Bitbucket Pipelines.

The remaining memory is then available to the service containers. After memory is allocated by the build container, 3,027 MB or 7,128 MB is left over for the service containers. By default, each service container can receive 1,024 MB or a custom amount between 128 MB and the maximum amount. This can be set by using the `memory` keyword in the service definition.

If your build step includes Bitbucket Pipes, it uses a built-in Docker service. By default, this Docker service occupies 1,024 MB of the build step's memory but can be configured to a custom amount by setting the memory between 128 MB to the maximum.

Now that we understand these limitations, let's learn how to define the services that can run on a given step.

How to do it...

The following is a set of examples of the types of services that can be defined within a build step. In this recipe, we're going to look at examples that use them:

- Database service

- Data store (for example, NoSQL) service

- Docker-in-Docker service

Let's look at how to define these services.

Defining a containerized database service

Let's start with what the definition would be for a database service that had all default values:

1. In the `definitions:` section, specify the service name in `services:`, add the Docker image using the `image:` keyword, and the necessary credentials (as secure variables) in the `variables:` section. This should look as follows:

```
definitions:
    services:
        mysql:
            image: mysql:5.7
            variables:
                MYSQL_DATABASE: test-db
                #set up password as secure variable and use here
                MYSQL_ROOT_PASSWORD: $password
```

2. To use the service in a step, add the service in the `services:` section of the step. This is shown in the following code snippet:

```
default:
    - step:
            services:
                - mysql
```

3. To customize the memory allocation, add the desired amount of memory, in MB, after the `memory:` keyword in the `definitions:` section. The code should now look as follows:

```
definitions:
    services:
        mysql:
            image: mysql:5.7
            memory: 2048 # double the mimimum
            variables:
```

```
MYSQL_DATABASE: test-db
#set up password as secure variable and use here
MYSQL_ROOT_PASSWORD: $password
```

We now have a database service running for our build step with 2,048 MB in its service container.

Defining a containerized data store service

Let's create our data store service in the same manner that we did in the previous example:

1. In the `definitions:` section, set up the service name in `services:` and add the Docker image using the `image:` keyword. These additions should look as follows:

    ```
    definitions:
        services:
            redis:
                image: redis:3.2
    ```

2. To use the service in a step, simply add it to the `services:` section of the step. We've also added a command we're running based on the service. This is shown in the following code snippet:

    ```
    default:
        - step:
            script:
                - redis-cli -h localhost ping
            services:
                - mysql
    ```

With that, we've learned how to define a containerized service and use it within the build step.

Now, let's learn how to invoke a Docker-in-Docker service.

Defining a service for Docker-in-Docker

Let's look at running a Docker service within our build step:

1. In the `definitions:` section, set up the service name in `services:`. This should look as follows:

    ```
    definitions:
        services:
            docker:
    ```

2. To use the service in a step, add it to the `services:` section of the step. We've also added a command we're running based on the service. This is shown in the following code snippet:

```
default:
  - step:
        script:
            - docker info
        services:
            - docker
```

3. You can give your Docker services custom names. Define the service with the custom name and set `type:` to `docker`. A detailed example of a custom Docker service with memory size customizations is shown in the following code snippet:

```
definitions:
    services:
        my-docker:
            memory: 5120
            type: docker

default:
    - step:
        services: my-docker
        size: 2x
        script:
            - docker info
```

With that, we've learned how to define Docker as a service from within Bitbucket Pipelines.

Next, we'll learn how to deploy our Bitbucket Pipelines output as a Docker image and push the image to a Docker repository. For that, we need to understand how to execute Docker commands. Let's see how that's done.

Using Docker commands in Bitbucket Pipelines

If you have a Dockerfile in your Bitbucket repository, you can use Bitbucket Pipelines to build the image and push it to your Docker repository. You can do this by executing Docker commands from within the `bitbucket-pipelines.yml` file. Let's take a closer look at how that's done.

Getting ready

Before adding Docker commands to the `bitbucket-pipelines.yml` file, we need to enable the following configurations:

- Allowing access to the Docker daemon
- Enabling Docker BuildKit

These configurations are part of `bitbucket-pipelines.yml`. Let's see where they go.

Enabling access to the Docker daemon

Access to the Docker daemon as a service can be done either by adding Docker as a service to an individual step, which is recommended so that you can keep track of how many services your overall pipeline is running, or by adding Docker as a service to all steps. Let's see how each alternative is done:

1. To add Docker as a service for the build step, make sure it is present in the `services:` section of the step. This is illustrated in the following code snippet:

    ```
    pipelines:
      default:
        - step:
            script:
              - ...
            services:
              - docker
    ```

2. To add Docker as a service globally, add `docker` in the `options:` section and set it to `true`, as shown in the following code snippet:

    ```
    options:
      docker: true
    ```

There are a few things to note about enabling access to the Docker daemon. Docker is defined as a service by default, so we don't need to define it in the `definitions:` section of `bitbucket-pipelines.yml`.

These days, creating Docker builds involves using Docker BuildKit. So, let's learn how to work with Docker BuildKit in Bitbucket Pipelines.

Enabling Docker BuildKit

Docker BuildKit is a default part of building when using Docker Desktop or Docker Engine v23.0 and beyond. We want to make sure it is enabled to ensure compatibility with these versions. So, let's explore how to work with Docker BuildKit.

To enable Docker BuildKit, make sure the DOCKER_BUILDKIT environment variable is set to 1, as shown in the following code snippet:

```
pipelines:
    default:
        - step:
          script:
              - export DOCKER_BUILDKIT=1
              - docker build .
          services:
              - docker
```

Now, let's learn how to use Docker commands in Bitbucket Pipelines.

How to do it...

With Docker and Docker BuildKit enabled, we can run most Docker commands. For security, reasons Bitbucket Pipelines has placed limitations on the Docker commands that can be run, as well as the modes of other Docker commands. A detailed list of the restrictions can be found at https://support.atlassian.com/bitbucket-cloud/docs/run-docker-commands-in-bitbucket-pipelines/.

In the meantime, we'll consider the common use cases for incorporating Docker into Bitbucket Pipelines. These include the following:

- Building a Docker image from a Dockerfile
- Passing secrets to Docker BuildKit from Bitbucket secured variables
- Passing secrets to Docker BuildKit from external secret managers
- Pushing a Docker image to a Docker repository

Let's examine these common use cases.

Building a Docker image from a Dockerfile

With Docker BuildKit enabled, Bitbucket Pipelines can build a Docker image. Let's look at this in detail:

1. Make sure that a Dockerfile exists at the root level of your Bitbucket repository.

    You may want to specify your image's name as a variable.

2. Add the following line in the script portion of your build step:

    ```
    - docker build -t $IMAGE_NAME .
    ```

With this done, let's move on to the next use case.

Passing secrets to Docker BuildKit with secured variables

If we need to pass secrets such as credentials or API keys to a BuildKit build, we can use secured variables in Bitbucket to pass them. Here's how we can do this:

1. Create the secure variable in Bitbucket.

    In the `bitbucket-pipelines.yml` file, add the `--secret` flag and set the ID to the secure variable. The code snippet shows `SECRET` as the secure variable:

    ```
    pipelines:
        default:
            - step:
                  name: 'BuildKit and secure variables'
                  script:
                  # Enable BuildKit
                  - export DOCKER_BUILDKIT=1
                  # Pass the secure variable into Docker build and
    prevent caching
                  - docker image build -t latest --secret id=SECRET
    --progress=plain --no-cache dockerfile
                  services:
                  - Docker
    ```

2. In the Dockerfile, add a `RUN` instruction that mounts the secure variable (using the `--mount=type=secret` flag) into the default Docker secret store (`/run/secrets/*`). This is illustrated in the following code snippet:

    ```
    FROM ubuntu:latest
    # Mount and print SECRET
    RUN --mount=type=secret, id=SECRET \
            cat /run/secrets/SECRET
    ```

There's another way secrets can be passed to Docker BuildKit: using an external secret manager. So, let's explore how to work with external secret managers.

Passing secrets to Docker BuildKit with external secret managers

We can also connect to external secret managers such as Hashicorp Vault or Google Cloud Secret Manager to pass secrets such as credentials or API keys to a BuildKit build. Let's learn how:

1. In the `bitbucket-pipelines.yml` file, get the secret from the manager, place the secret in a pipeline file, add the `--secret` flag, and identify the source as the pipeline file. Remember that the pipeline file will be deleted when the pipeline step is complete and the container is removed. The following code snippet shows `SECRET` as the secure variable:

    ```
    pipelines:
        default:
    ```

```
        - step:
            name: 'BuildKit and external secret managers'
            script:
            # Enable BuildKit
            - export DOCKER_BUILDKIT=1
            # This is where the call to the external secret
manager resides.  We assume here that it has added the secret to
"/secret_file"
            # Pass the secure variable into Docker build and
prevent caching
            - docker image build -t latest --secret
id=SECRET,src=/secret_file --progress=plain --no-cache
dockerfile
            services:
            - docker
```

2. In the Dockerfile, add a RUN instruction that mounts the secure variable (using the `--mount=type=secret` flag), including the pipeline file containing the secret, into the default Docker secret store (`/run/secrets/*`). This is illustrated in the following code snippet:

```
FROM ubuntu:latest
# Mount and print SECRET
RUN --mount=type=secret, id=SECRET,dst=/secret_file \
        cat /run/secrets/SECRET
```

With that, we've built an image that passed a secret found in an external source. Now, let's look at the various actions you can perform, including pushing to a Docker registry.

Pushing a Docker image to a Docker registry

You can push images you create to Docker Hub or another registry as part of a Bitbucket Pipelines script execution. Follow these steps:

1. Create variables for the image name, username for the Docker repository, and password for the Docker repository. The username and password should be secure variables.

2. Add commands to the script portion of the step to deploy the Docker image. To do this, you must do the following:

    I.   Log in to the Docker repository.

    II.   Push the image via `docker push`.

    The following code snippet shows a basic example of how to build and push a Docker image:

```
- step:
    name: Build
    script:
```

```
            # Build the Docker image (assumes the Dockerfile is at the
    root level of the repository)
            - docker build -t $IMAGE_NAME .
            # Authenticate with the Docker registry (this example is
    Docker Hub)
            - docker login --username $DOCKER_HUB_USERNAME --password
    $DOCKER_HUB_PASSWORD
            # Push the image to the Docker registry
            - docker push $IMAGE_NAME
        services:
            - docker
```

With that, we've learned how to deploy our application as a Docker image by building it and pushing the resulting image to a Docker registry. We can take our example a step further by deploying the Docker image to a Kubernetes cluster. We'll do this in the next section.

Deploying a Docker image to Kubernetes using Bitbucket Pipelines

A potential next step after building the Docker image is to deploy it to a Kubernetes cluster. By taking advantage of the redundancy capabilities in a cluster, we can perform application upgrades without a service outage occurring.

Getting ready

The instructions in this recipe assume that you have an existing Kubernetes cluster or minikube environment that was created manually.

In addition, you must manually define a deployment that runs the application in Kubernetes. So, let's learn how to create a deployment.

Ensure the application name and the Docker registry username are available for easy reference. Execute the kubectl command, including the necessary flags, as shown in the following code snippet. These will include the application name and Docker registry name as part of the image name:

```
kubectl run <my.app> --labels="app=<my.app>" --image=<my.dockerhub.
username>/<my.app>:latest --replicas=2 --port=8080
```

Now that we have established a deployment in Kubernetes, let's learn about the continuous deployment aspect of the application when using Bitbucket Pipelines.

How to do it...

We can integrate with the `kubectl` application from Bitbucket Pipelines in one of two ways:

- Using a Pipe to integrate
- Setting up a service

Let's take a look at each method in more detail.

Executing kubectl using pipes

As we saw in *Chapter 6*, Bitbucket Pipes are pre-packaged integrations with common third-party tools and utilities. They can be easily added into Bitbucket Pipelines steps and will execute through separate containers as services.

Let's learn how to incorporate the pipe for `kubectl`:

1. Set up the `kubeconfig` file for reading. This file needs to be `Base64` encoded and then stored as a secure variable. You can use the following code to do so:

   ```
   KUBE_CONFIG_BASE64=$(cat ~/.kube/config | base64)
   ```

2. Add the pipe to the script portion of the step. The pipe definition should look like this:

   ```
   - step:
     name: Deploy
     deployment: production
     script:
        -pipe: atlassian/kubectl-run:1.1.2
         variables:
            KUBE_CONFIG: $KUBE_CONFIG
            KUBECTL_COMMAND: 'apply'
            RESOURCE_PATH: 'deployment.yml'
   ```

With that, we have deployed to Kubernetes using a pipe via Bitbucket Pipelines.

Sometimes, we need to execute a different version of `kubectl` than what's provided on a Pipe. In this case, executing from the Atlassian-provided Docker image for `kubectl` is a better alternative. One reason for incorporating a different version may be to ensure compatibility with an existing Kubernetes cluster on a legacy version. Let's explore that option now.

Executing kubectl using a kubectl Docker image

Bitbucket Pipelines also has a version of `kubectl` that's encapsulated in its own Docker image. This image is located on Docker Hub at `https://hub.docker.com/r/atlassian/pipelines-kubectl` and can be used within a Bitbucket Pipelines script to execute `kubectl` commands.

To deploy on Kubernetes using a `kubectl` service, perform the following steps:

1. Within the deployment step, define the Docker image using the `image:` keyword.

2. Set up our `kubeconfig` file. This time, we're decoding Base64 to create a temporary file that gets destroyed after execution:

    ```
    echo $KUBECONFIG | base64 -d > kubeconfig.yml
    ```

3. Execute the `kubectl` command to apply a new version of the application:

    ```
    - kubectl --kubeconfig=kubeconfig.yml apply -f deployment.yml
    ```

4. Putting this all together, we have the following code snippet:

    ```
    -step:
        name: Deploy to Kubernetes
        image: atlassian/pipelines-kubectl
        script:
            - echo $KUBECONFIG | base64 -d > kubeconfig.yml
            # Run deployment command using kubectl
            - kubectl --kubeconfig=kubeconfig.yml apply -f deployment.
    yml
    ```

    As seen in the *Configuring deployments* recipe in *Chapter 8*, you can monitor your deployment using the deployment dashboard.

With that, you've learned how to build our application as a Docker image and deploy it to a Kubernetes environment.

Our last stop in examining how we can leverage Docker moves us from pipelines to runners. So, let's learn how to configure Docker-based runners on Linux.

Setting up Docker-based runners on Linux

This application of this self-hosted runner allows for the ultimate in dynamic configuration. By allowing runners inside Docker containers, we can add or subtract runners as needed.

We'll start with a Linux environment, install Docker, and load and run the Docker image for the Bitbucket runner software. Let's take a look at the complete picture of setting up Docker-based runners.

Getting ready

Our Linux environment has some prerequisites we must cover before we can proceed. First, we need to understand our Linux environment. This environment should have the following features:

- You should be using the 64-bit version of Linux.
- A minimum of 8 GB of RAM should be allocated to the host for the runner. If you know that you are going to need a lot of room (for example, due to more build steps), you should allocate more memory.
- At least 512 MB must be allocated for the runner container.
- Docker v19.03 or above must be installed.

With the Linux environment set up in this manner, we have to look at best practices for our Linux environment recommended by Atlassian. Atlassian recommends the following environment configurations:

- Disabling **swap** space in your Linux environment
- Configuring vm.swappiness

Let's take a closer look at these recommendations.

Disabling swap space

Depending on the Linux distribution you're using, you may not have the necessary commands installed. If the following commands aren't available in your Linux environment, you can install them using the distribution's preferred package manager:

1. Check if swap is enabled:

   ```
   sudo swapon -sv
   ```

   Existing swap partitions will appear, as shown in the following output, if swap is enabled:

   ```
   NAME        TYPE       SIZE    USED PRIO
   /dev/sda3 partition   2G 655.2M   -1
   ```

2. Disable swap by executing the following command:

   ```
   sudo swapoff -av
   ```

3. Remove any configured swap partitions on /etc/fstab.
4. Reboot your Linux machine.
5. Repeat these steps until no swap partitions appear.

At this point, we have eliminated one source of swap storage. However, we should eliminate other sources. For that, we will take a look at configuring vm.swappiness.

Configuring vm.swappiness

Again, some Linux distributions may not have the commands specified in the following steps. If this is the case, install the required commands using the Linux distribution's recommended package manager.

Let's take a close look at correctly configuring vm.swappiness to disable swap:

1. Check the value of vm.swappiness using the following command:

    ```
    sudo sysctl -n vm.swappiness
    ```

2. If the value isn't 1, configure vm.swappiness by performing the following steps:

    I.   Open /etc/sysctl.conf and add vm.swappiness=1 to its own line in the file.

    II.   Save your changes.

    III. Reboot the Linux machine.

3. If subsequent examinations of the value of vm.swappiness are anything other than 1, repeat these steps and ensure the setting is configured correctly in /etc/sysctl.conf.

The next step is highly recommended to maintain the proper operation of your Linux environment. Periodically, stale Docker container images should be cleaned up. Let's see how to schedule this operation.

Automating the cleanup process for stale Docker images

Our Linux environment should regularly remove unused Docker images to save on disk space. We want to ensure we have adequate disk space in our Linux environment so that we can continue to operate our runner and ensure its availability for Bitbucket Pipelines jobs. The command to remove unused Docker images is docker system prune -af. A common way of scheduling an automated means for running a command is by using **cron**. Let's see how that can be done:

1. For the correct user, open their crontab file by typing the following command:

    ```
    crontab -e
    ```

    Append the command to the crontab file while setting up the correct frequency, dates, and times. The following example runs the command on Sundays at midnight:

    ```
    0 0 * * 0 docker system prune -af
    ```

2. Save the file and exit the editor.

With that, we've used cron to automate the deletion of unused Docker images. Now, it's time to set up our runner.

How to do it...

With the preliminary steps out of the way, it's time to connect our Linux environment to Bitbucket so that it can be used as a runner. Let's examine how to do that:

1. On Bitbucket, define a new runner. A workplace runner can be defined by clicking the administration cog at the top-right corner of the screen and selecting **Workspace settings**:

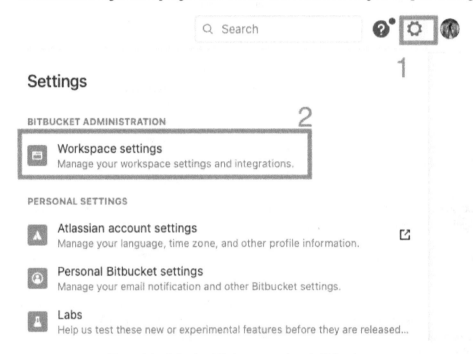

Figure 9.1 – Selecting Workspace settings in Bitbucket

2.   In the menu bar on the left, select **Workspace runners**:

Figure 9.2 – Selecting Workspace runners

Alternatively, if you're setting up a runner for use within a repository, select the repository and select **Repository settings** in the repository's sidebar.

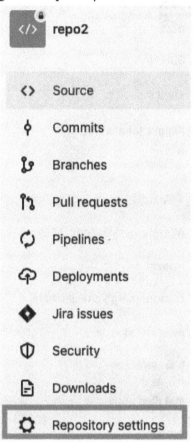

Figure 9.3 – Selecting Repository settings

3.   In the **Repository settings** sidebar, select **Runners** under **Pipelines**.

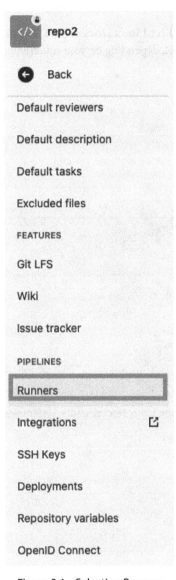

Figure 9.4 – Selecting Runners

4.   Regardless of whether you're selecting the runner for the workspace or the repository, you can create the runner by selecting the **Add runner** button.

In the modal that appears, select **Linux Docker (x86_64)** or **Linux Docker (arm64)** in the **System and architecture** panel, depending on your underlying hardware platform. Click **Next**.

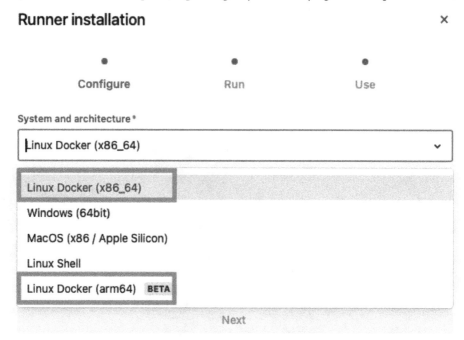

Figure 9.5 – Selecting Linux Docker under System and architecture

In the next modal, copy the Docker command that's displayed and paste it into a Terminal window in your Linux environment. This command goes to the Docker registry to retrieve the Bitbucket Pipelines runner software as a Docker image and creates the container.

Runner installation   ✕

Configure   Run   Use

Run the command below to install the runner. This token will not be displayed again.

By installing the runner, I agree to the **Atlassian Software License Agreement** and acknowledge the **Privacy Policy**.

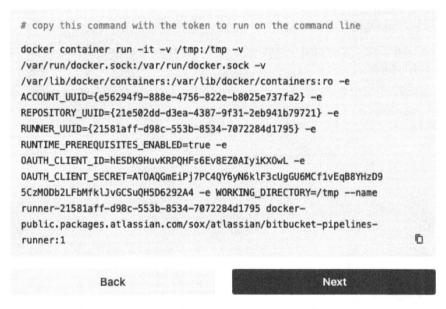

```
# copy this command with the token to run on the command line

docker container run -it -v /tmp:/tmp -v
/var/run/docker.sock:/var/run/docker.sock -v
/var/lib/docker/containers:/var/lib/docker/containers:ro -e
ACCOUNT_UUID={e56294f9-888e-4756-822e-b8025e737fa2} -e
REPOSITORY_UUID={21e502dd-d3ea-4387-9f31-2eb941b79721} -e
RUNNER_UUID={21581aff-d98c-553b-8534-7072284d1795} -e
RUNTIME_PREREQUISITES_ENABLED=true -e
OAUTH_CLIENT_ID=hESDK9HuvKRPQHFs6Ev8EZ0AIyiKXOwL -e
OAUTH_CLIENT_SECRET=ATOAQGmEiPj7PC4QY6yN6klF3cUgGU6MCf1vEqB8YHzD9
5CzMODb2LFbMfklJvGCSuQH5D6292A4 -e WORKING_DIRECTORY=/tmp --name
runner-21581aff-d98c-553b-8534-7072284d1795 docker-
public.packages.atlassian.com/sox/atlassian/bitbucket-pipelines-
runner:1
```

Back   Next

Figure 9.6 – Copying the Docker command to pull the runner

5. You may want to get the most up-to-date version of the runner when you start the runner again or just to make sure you have the latest one. To perform this update, execute the following `docker pull` command in your Linux environment:

```
docker image pull docker-public.packages.atlassian.com/sox/
atlassian/bitbucket-pipelines-runner:1
```

You may encounter the following error when starting the runner:

```
docker: Error response from daemon: docker: Error response
from daemon: Conflict. The container name "/runner-76b247e7-
b925-5e7b-9da2-1cda14c4ff2c" is already in use by container
"c3403236e3af5962ed3a9b8771561bd2021974941cc8a89a40c6c-
66cecb18f53". You have to remove (or rename) that container to
be able to reuse that name.
See 'docker run --help'.
```

If this is the case, remove the runner by executing the following command:

```
docker container rm -f runner-76b247e7-b925-5e7b-9da2-
1cda14c4ff2c
```

6. It's possible to change the working directory that's used by the runner on your Linux machine. To change the working directory, add the following two flags to the docker run command:

- The -v flag and the directory, as seen on the host machine.

- The -e flag and the desired mount point inside the runner. You can use any desired mount point, but it must match the value of the WORKING_DIRECTORY environment variable.

An example of using the docker run command is shown in the following code snippet:

```
docker run [all existing parameters] -v /mydir:/mydir -e
WORKING_DIRECTORY=/mydir
```

With that, we've established Docker-based runners for any future executions of Bitbucket Pipelines.

Part 3: Maintaining Operations

After release, the focus turns to ensuring that the environment, with its new features and products, maintains the same level of performance, scalability, and security as before. Measurements of performance, both in the context of how the system is operating and whether it is delivering its promised value, are taken and displayed.

The displays reflect a focus on observability, ensuring not only that the metrics exist but that they are available and visible to everyone: developers, operations people, site reliability engineers, and others in the business.

When problems occur, people from these disciplines come together to collaborate on the problem and find a solution.

In this part, we will explore how Atlassian tools such as Jira, Opsgenie, and Compass work together and with other tools to allow observability to all disciplines and the rapid escalation and resolution of problems.

This part has the following chapters:

- *Chapter 10, Collaborating with Operations through Continuous Deployment and Observability*
- *Chapter 11, Monitoring Component Activity and Metrics Through CheckOps in Compass*
- *Chapter 12, Escalating Using Opsgenie Alerts*

Collaborating with Operations through Continuous Deployment and Observability

In this chapter, we will close the circle of DevOps by looking at the integrations to tools used primarily by operations. Developers using Jira can find the results of deployments by integrating Jira with **Continuous Deployment** (**CD**) tools, such as Bitbucket, GitLab, GitHub, or Jenkins. Observability tools that monitor for performance, such as Dynatrace and Datadog, can link to Jira issues through integrations.

This chapter contains the following recipes:

- Connect Jira with continuous deployment tools
- Connect Jira with observability tools

Technical requirements

To complete this chapter, you will need the following:

- Jira
- A GitLab account (https://about.gitlab.com/)
- A Datadog account (https://www.datadoghq.com/)

Connect Jira with continuous deployment tools

In this recipe, we are going to learn how to integrate Jira with CD tools to view deployment status. In *Chapter 4*, we learned that **Continuous Integration** (**CI**) is the practice of making incremental improvements or changes to a code base and automating the testing and validation of those code changes.

CD can be considered an extension of CI that handles the automated infrastructure provisioning and application release process.

The objectives for this recipe are as follows:

- Associate a GitLab repository with a Jira project
- Add the Jira integration to the GitLab SaaS application
- Perform updates to the code base, execute a merge request, and deploying the code
- See the GitLab deployment reflected in the Jira project and issue

Getting ready

To execute this recipe, you will need the following:

- Jira administration permissions in order to add the GitLab application
- A GitLab account and repository with actual or sample code

How to do it...

We will use the following steps to achieve this recipe's objectives:

1. First, we need to ensure that **Deployments** is enabled for the Jira project you are working on.

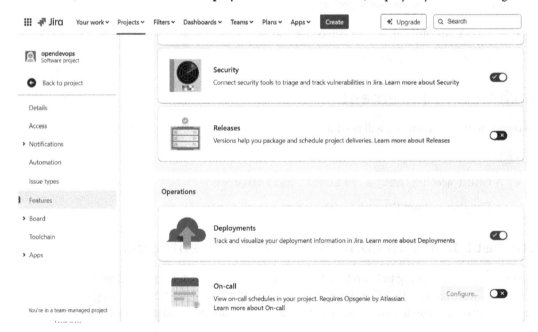

Figure 10.1 – Enable Deployments for a Jira project

2.   Next, we need to verify the GitLab and Jira integration. We covered the integration of GitLab and Jira in *Chapter 4*. Please refer back to that chapter if necessary.

Once the GitLab integration has been verified, we will need to connect a GitLab repository to a Jira project.

3.   In your desired Jira project, navigate to **Project settings | Toolchain**.

Then select **Add tool**.

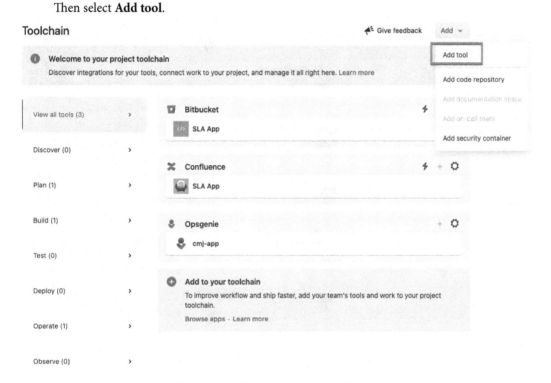

Figure 10.2 – Add to the project toolchain

You will see the **GitLab for Jira Cloud** option available from the list of recommended tools.

4.   Select **Add to project**.

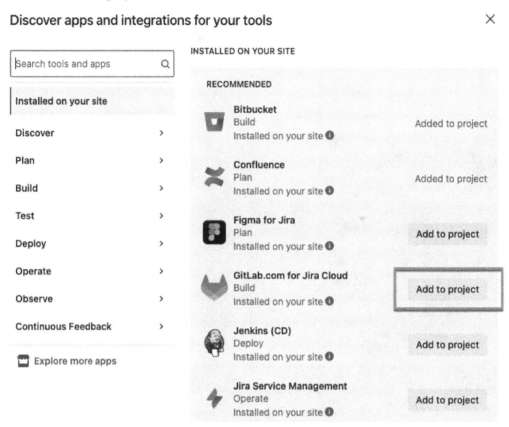

Figure 10.3 – Add GitLab to Jira project

> **Important note**
>
> The GitLab box is added to the list of existing tools for the project. Now we need to associate a repository from GitLab with the project, and with the issues within the project related to working on the code base.

5. Select the + **Add code repository** option.

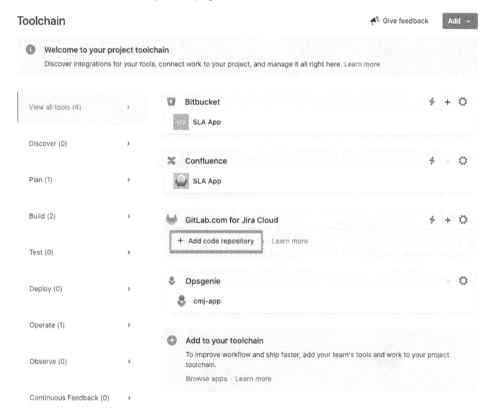

Figure 10.4 – Add GitLab code repository

Since GitLab is already integrated with Jira, a pop-up window is displayed with a drop-down option. Select the dropdown and view the available GitLab repositories. You may select one or more repositories to associate with a Jira project.

6. In this example, we are selecting a repository associated with our mobile application code base.

After selecting a repository, click the **Add repository** button.

Figure 10.5 – Select GitLab repository

The GitLab box in the toolchain now reflects the selected repositories.

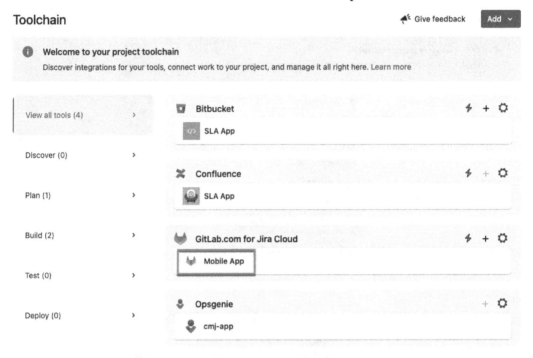

Figure 10.6 – GitLab repository added

> **Tip**
>
> Now that we have the Jira project associated with a GitLab repository, we need to look at the issues within the project. We need to identify or create a new project to associate with code changes and deployment.
>
> When we make changes to the code base in GitLab, we will need the issue's key to add in the branches and commit comments. This will tell GitLab which issues are associated with the deployment. This is called Smart Commits.

In the following Jira example, we have an issue with a key value equal to **ACR-6**. The issue is currently in the **IN PROGRESS** status column.

Figure 10.7 – Jira Issue Key

7. Now let's go back over to GitLab and perform some updates to the code base in relation to the ACR-6 issue. First we'll make a new branch for the specific changes. In GitLab, under the **Code** menu option, select **Branches**.

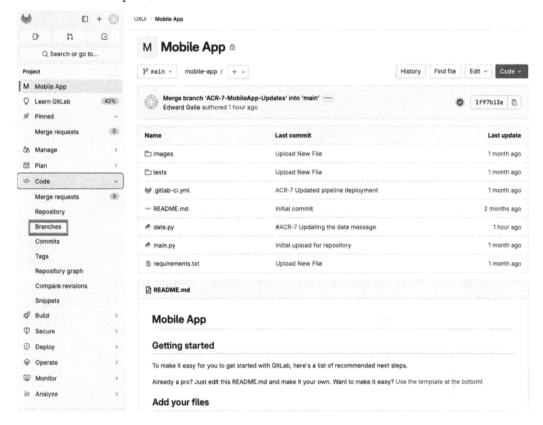

Figure 10.8 – GitLab code repository branch

This will give you the option to create a new branch.

8. In the **Branch name** field, make sure to use the ACR-6 issue key so we can tie this change back to the issue in Jira. Then select **Create branch**.

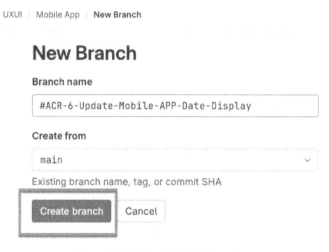

Figure 10.9 – Create GitLab branch

9. From the new branch, we will select some code to modify. In this example, we are going to make some simple updates to a Python script. There are several ways to perform these code modifications. To keep it simple, we will open the code in the Web IDE.

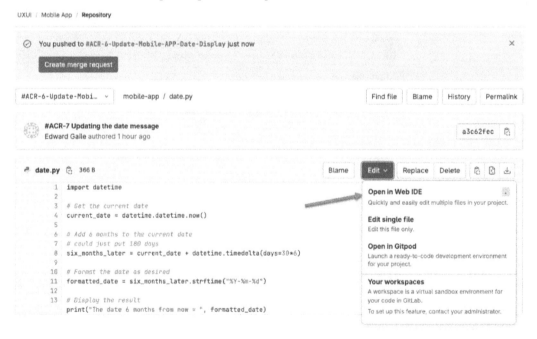

Figure 10.10 – Open in Web IDE to modify code

We have updated the code to display the date four months from now.

10. Once our changes are complete, we can click on the source control button and commit the changes back to our new branches. In the commit message, it is important to use the issue key again (ARC-6) to tie this commit back to the associated Jira issue.

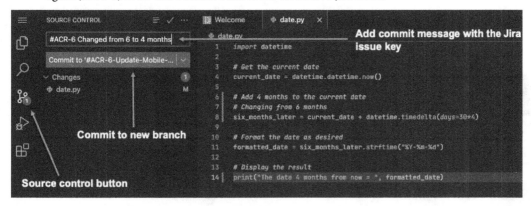

Figure 10.11 – Modify then commit code

11. After committing the changes, we can go back to GitLab and then execute a merge request to pull the code back into the main branch and deploy the code.

Select **Create merge request**.

Figure 10.12 – Create merge request

Once the merge request has been created, you will see the pipeline process running. The pipeline defined for this example is to perform a build, run two test jobs, and finally deploy the code to a production environment.

The code can then be merged into the main branch and the merge request closed out.

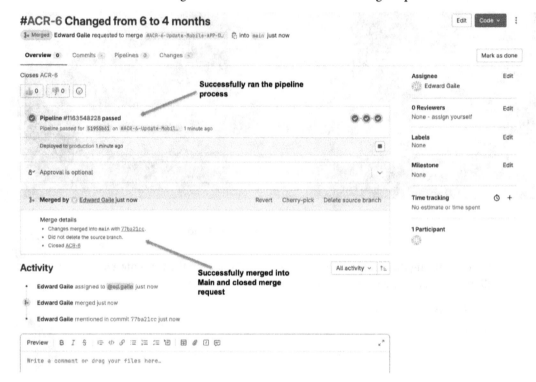

Figure 10.13 – Code merge

12. After the merge request and deployment are complete, we can switch back over to the Jira project and see that the issue was automatically transitioned to a done status.

We can also select the **Deployments** menu option in the project's left-side panel to see all deployments.

Figure 10.14 – Jira deployment actions

We can see the deployment for ACR-6 that was just executed displayed in the **Deployments** view.

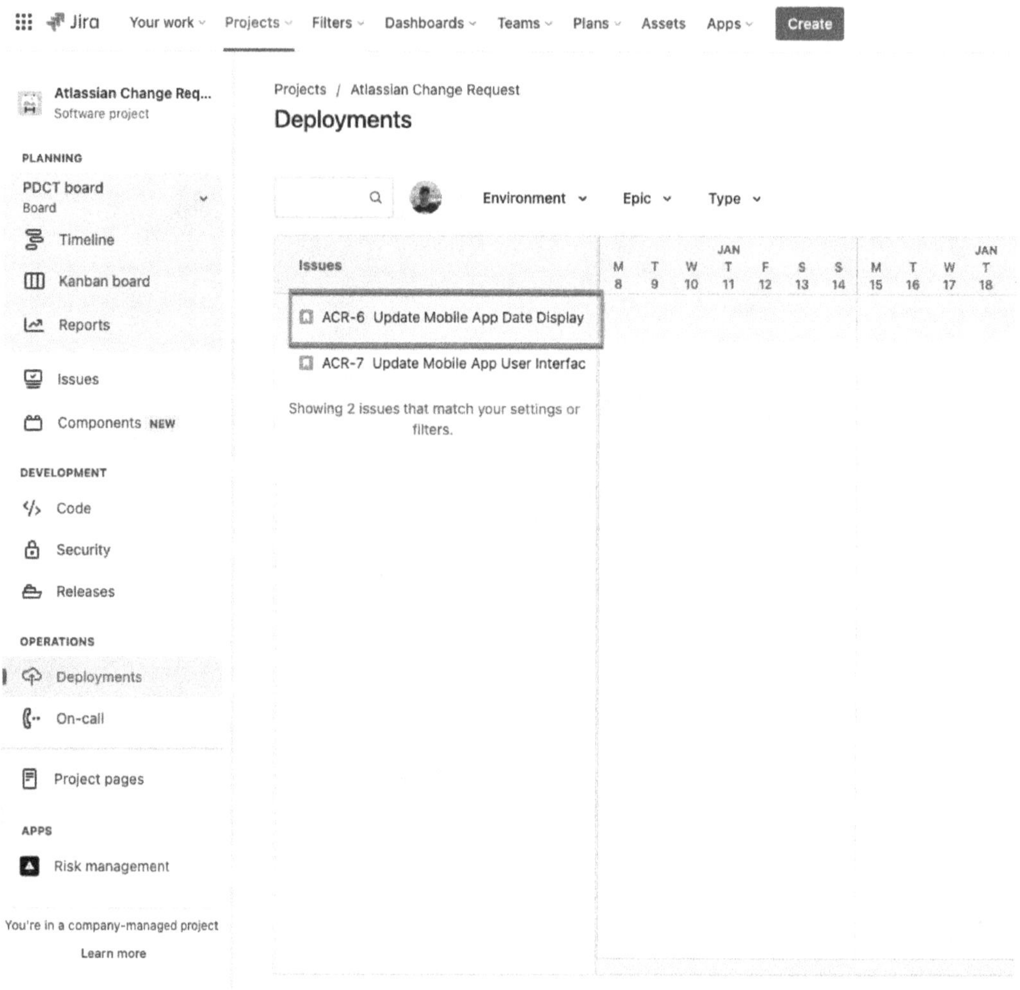

Figure 10.15 – Jira Deployments view

If we select and view the actual ACR-6 issue, we can see all of the associated GitLab activities. The commits and merge requests are displayed under the **Web links** section. Clicking on any of those links will take you to the corresponding artifact in GitLab.

> **Tip**
> You can also view the **Development** panel and see all of the associated commits, builds, and releases.

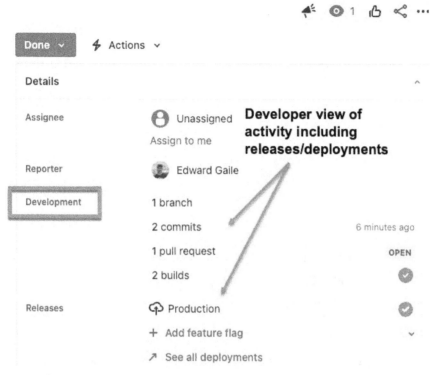

Figure 10.16 – Jira issue Developer view

You have now successfully integrated a CD app (GitLab) with Jira. This will add to the overall visibility of tracking your code all the way through a release.

Connect Jira with observability tools

DevOps observability is a practice within DevOps that focuses on gaining comprehensive insights into the behavior, performance, and health of applications and infrastructure. Observability is achieved by collecting, aggregating, and analyzing data from various sources within a system, including logs, metrics, and traces. In this section, we will learn how to connect Jira with logging and monitoring tools to correlate Jira issues with problems.

This recipe has the following objectives:

- Configuring a Datadog account with a Jira instance
- Demonstrating a Jira issue created from Datadog

Getting ready

To execute this recipe, you will need the following:

- Jira administration permissions in order to add the GitLab application
- A Datadog account

How to do it...

To implement the Datadog integration, we need to perform the following steps:

1. Install the Jira integration in Datadog.
2. Create an application link to Datadog from Jira.
3. Connect the Jira instance.
4. Create a webhook in Jira (if using Datadog Case Management).
5. Add a project to Datadog Case Management.
6. Create a case and see the reciprocal Jira issue created.

What about a Datadog app for Jira?

Datadog does not have an app that can be installed into Jira, even though the Jira Marketplace will show a listing available when searched. Let's use the following steps to find and install it:

1. Navigate to the **Discover apps and integrations for Jira** page and search for `datadog`.
2. Click on the **Datadog Jira Integration** option.

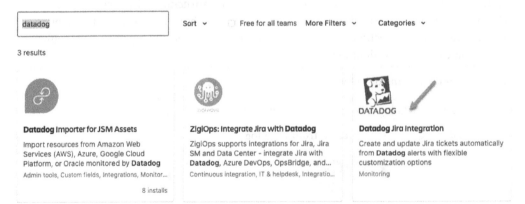

Figure 10.17 – Datadog Jira integration

3. The Datadog **Overview** page is displayed. Click on **Get app**.

Figure 10.18 – Datadog Overview page

This opens a pop-up window indicating that the integration configuration needs to be done on the Datadog side.

Installation instructions

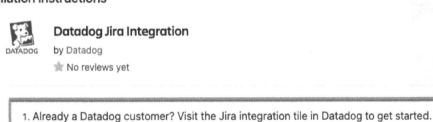

Figure 10.19 – Integration configuration starts in Datadog

Now, in order to configure the Datadog integration for Jira, you need to work from your Datadog account.

Installing the Jira integration in Datadog

We will use the following steps to install the Jira integration in Datadog:

1. Switching over to your Datadog account, select the **Integrations** option in the left-hand menu panel. This will bring you to the **Integrations** page.

2. Type `jira` into the search box and press *Enter*. The Jira integration tile will be displayed.

3. Select the Jira integration tile.

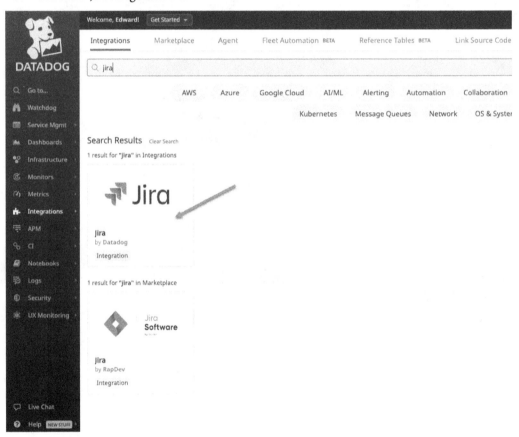

Figure 10.20 – Datadog Jira integration tile

4.   A **Jira Integration** pop-up window will be displayed. Select the **Configuration** tab.

Jira Integration

Jira is an issue and project tracking system for software teams.

`AVAILABLE`

Overview | Configuration |

Integrate with Jira to:

- Create and update issues using @jira mentions
- Receive a Datadog event each time a new Jira issue is opened

Figure 10.21 – Jira integration configuration

The **Configuration** page will be displayed, with the ability to configure a new Jira account. As the instructions indicate, an application link for Datadog will need to be created in Jira. One of the items needed for the integration link is the provided public key. Go ahead and copy this key to your clipboard.

Jira Integration

Jira is an issue and project tracking system for software teams.

`✓ INSTALLED`

Overview   Configuration

Accounts

> appfire-demos-sandbox-775.atlassian.net

∨ New Account

| Create an Application Link |

- Application links allow Datadog to securely access Jira.
- Create an application link and enter the information below to connect your Jira account to Datadog.
- For instructions on how to create an application link for Datadog, visit our documentation. ☑

Public key:

MIIBIjANBgkqhkiG9w0BAQEFAAOCAQ8AMIIBCgKCAQEAvCZf9CBIW0KZAjIgk9z4hNP/bz3/RQw5J80Pnw4+D/BnIji5Mn351VXm777oR78U4J
+8UATENyMPwuoZnqLc5poVscxvQ8Ns6VD4ge6k6Q2EiMhu5Sn2icx5oxnTtgn1anfgoZlazI8RkdcKkWIo6JJG6eMeEqrVu85Pz/gNMfRQbHlA
8aYRtmFFJNgEuxq261fpJFZJyZd89iV/9NtH08+EdnyIf34hYyaNpOxr1A8UgVQcU1jUxZvmQt/u28HocdFPHVB8Q9RYP+xinBA0iG6yzmGKaa
QsjCGczb6ZK84KMZM5Gxe+/G0nBFXSb4nukPr5HvLimDWN4Jl4F8m6kQIDAQAB

Authorize Datadog

Jira URL: | Enter Jira instance URL |

Consumer Key: | Enter consumer key |

Cancel   **Connect**

Figure 10.22 – Jira Integration account information

5. Switching over to your Jira instance, we now need to navigate to the **Products** menu and add
    an application link. Select the gear icon in the top-right corner and click on **Products**.

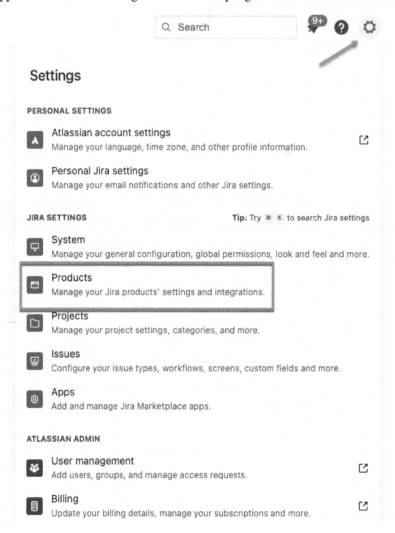

Figure 10.23 – Jira Product selection

6. From the **Product** page, scroll down the left-hand menu panel and select **Application links**.

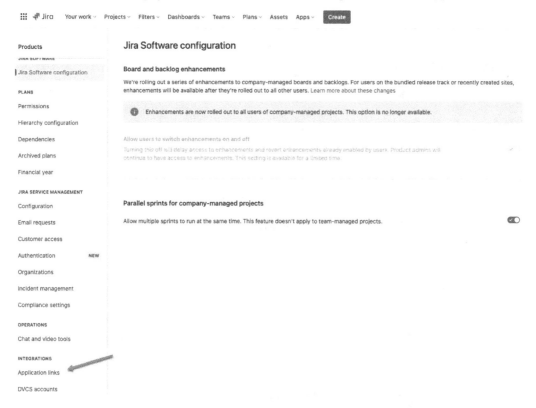

Figure 10.24 – Jira Application links

7. Now we can add a new application link by selecting **Create link**.

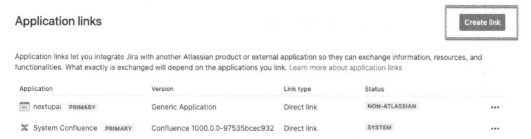

Figure 10.25 – Create application link

8.   Now we can begin to configure the Datadog application link:

I.   Select the **Direct application link** option.

II.   Add the Datadog application URL of `app.datadoghq.com`.

III.   Select **Continue**.

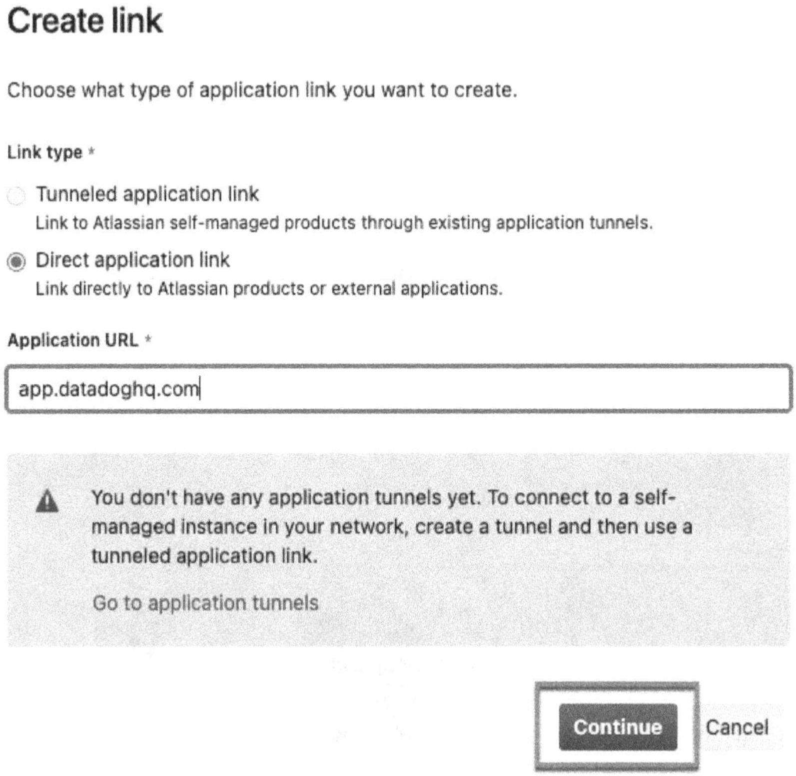

Create link

Choose what type of application link you want to create.

Link type *

○ Tunneled application link
   Link to Atlassian self-managed products through existing application tunnels.

 Direct application link
   Link directly to Atlassian products or external applications.

Application URL *

| app.datadoghq.com |

> ⚠ You don't have any application tunnels yet. To connect to a self-managed instance in your network, create a tunnel and then use a tunneled application link.
>
> Go to application tunnels

 Cancel

Figure 10.26 – Application link configuration

9. Ignore the **Configure Application URL** warning and click the **Continue** button.

Configure Application URL

⚠ The supplied Application URL has redirected once. Please check the redirected URL field to ensure this is a URL that you trust.

Entered URL

http://app.datadoghq.com

☐ Use this URL

Checking this will skip any further validation and use this URL for the link.

New URL*

https://app.datadoghq.com/

Continue   Cancel

Figure 10.27 – Application URL configuration

10. For the application link, continue by filling out the following:

I.   **Application Name** should be DataDog.

II. For **Application Type**, choose **Generic Application**.

III. Check the **Create incoming link** box.

IV.   Click **Continue**.

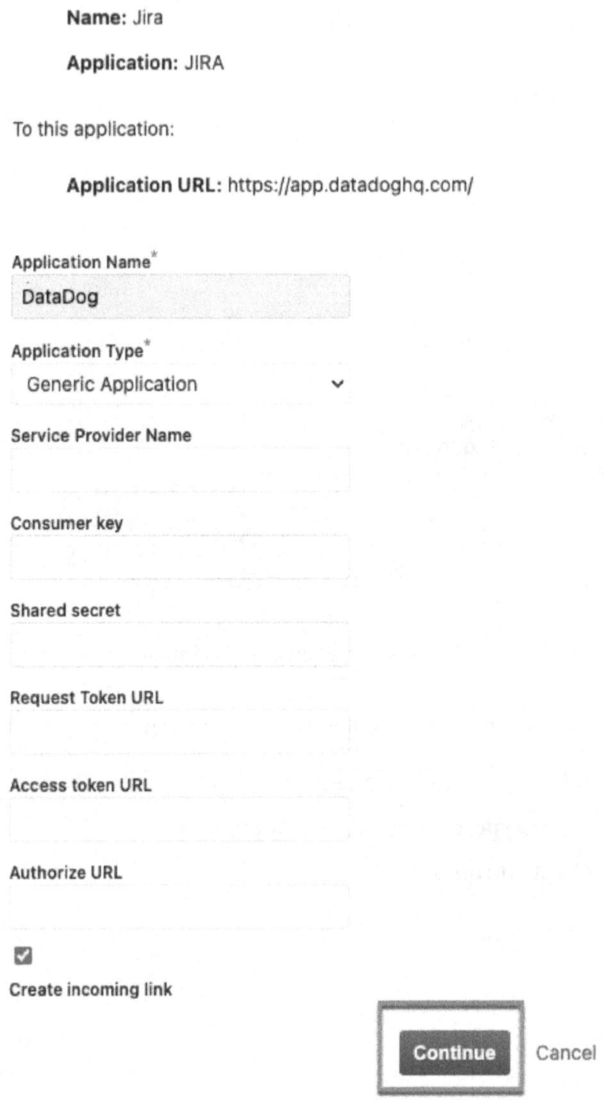

> **Name:** Jira
>
> **Application:** JIRA
>
> To this application:
>
> **Application URL:** https://app.datadoghq.com/
>
> Application Name[*]
>
> DataDog
>
> Application Type[*]
>
> Generic Application   ⌄
>
> Service Provider Name
>
> Consumer key
>
> Shared secret
>
> Request Token URL
>
> Access token URL
>
> Authorize URL
>
> ☑
>
> Create incoming link
>
> Continue   Cancel

Figure 10.28 – Details needed for Application link configuration

11. For the final part of the configuration process, fill out the following:

I.   **Customer Key** should be user-defined (i.e., `datadog-jira`). You will need to use this value to finish the Datadog Jira integration.

II.   **Consumer Name** is `Datadog`.

III. For **Public Key**, paste in the public key copied to your clipboard from the Datadog Jira configuration step.

IV.   Click **Continue**.

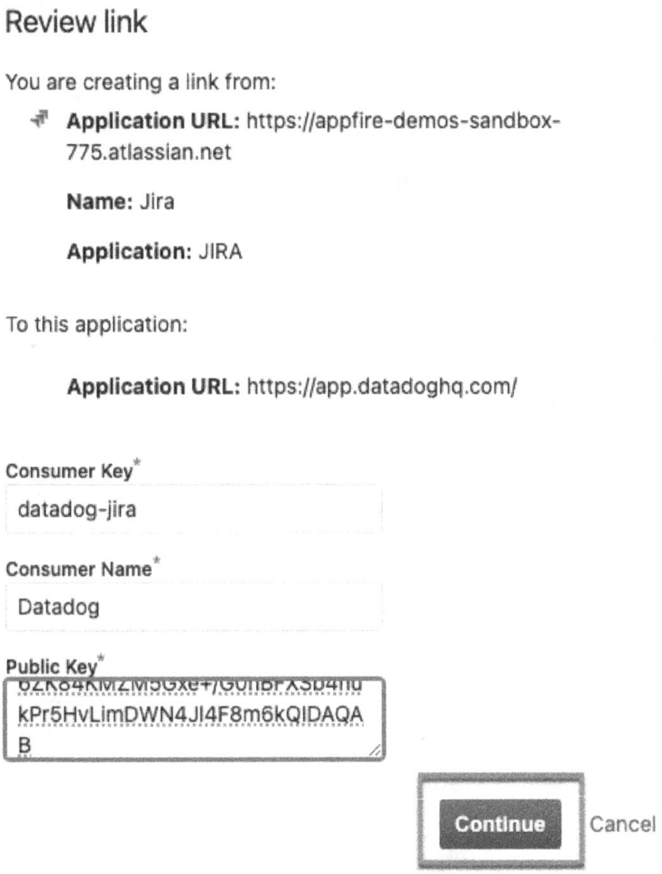

Review link

You are creating a link from:

⚓ **Application URL:** https://appfire-demos-sandbox-775.atlassian.net

Name: Jira

Application: JIRA

To this application:

Application URL: https://app.datadoghq.com/

Consumer Key*

datadog-jira

Consumer Name*

Datadog

Public Key*

oZK84KMZM9Gxe+/GUllbFXSb4IIU
kPr5HvLimDWN4JI4F8m6kQIDAQA
B

[Continue] Cancel

Figure 10.29 – Finalizing the Application link configuration

The new Datadog application link will be displayed and ready for use.

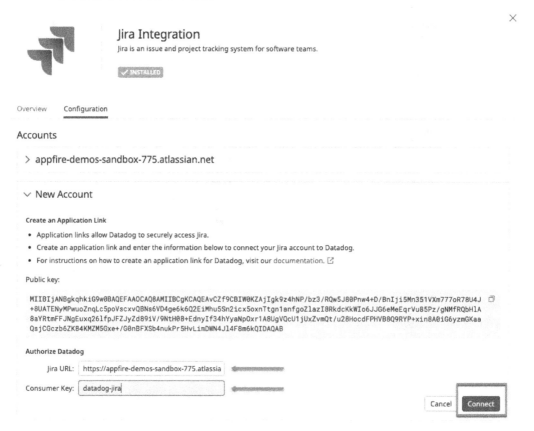

Figure 10.30 – Application link created

12. Switch back to the Datadog application to continue the Jira integration configuration. We now need to add the Jira Cloud URL and the consumer key we created for the Datadog application link in Jira. Then select **Connect**.

Figure 10.31 – Datadog Jira connection

Important note

Now that the Jira instance has been connected, we will have the option to set up a webhook for **bi-directional integration** with Datadog Case Management and an issue template to create Jira issues from alerts. In this example, we are going to configure the **Datadog Case Management** option.

13. You will need to copy the provided webhook URL to your clipboard.

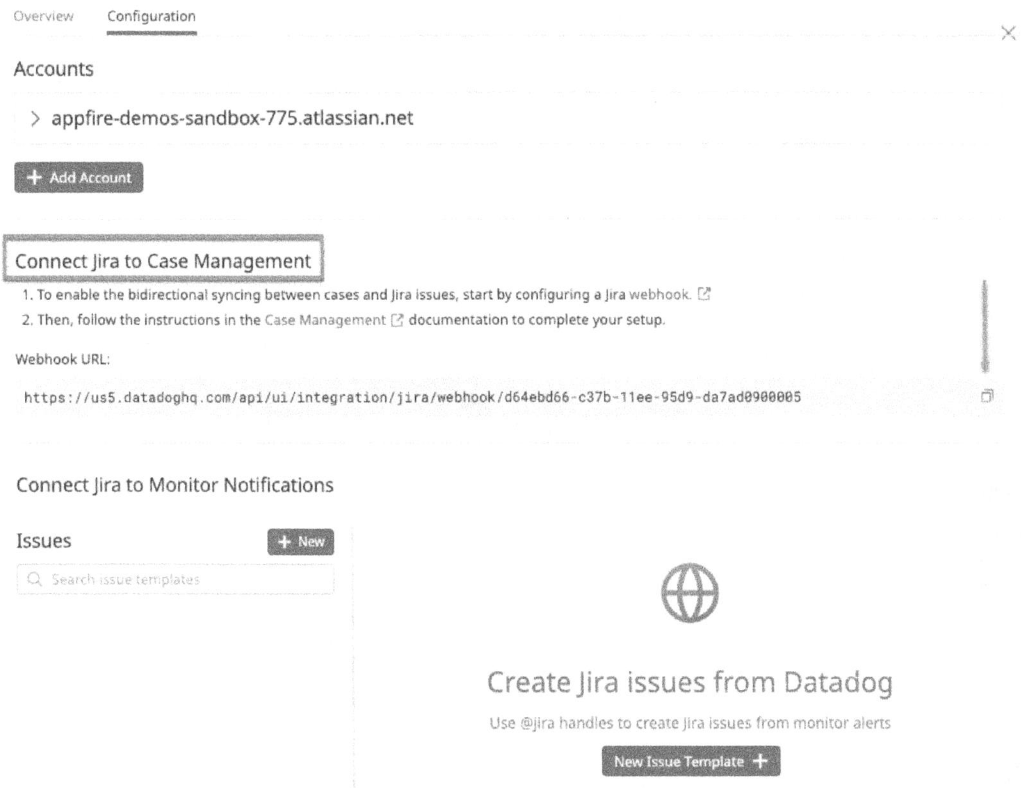

Overview   Configuration   ✕

Accounts

> appfire-demos-sandbox-775.atlassian.net

➕ Add Account

Connect Jira to Case Management

1. To enable the bidirectional syncing between cases and Jira issues, start by configuring a Jira webhook. ☐
2. Then, follow the instructions in the Case Management ☐ documentation to complete your setup.

Webhook URL:

https://us5.datadoghq.com/api/ui/integration/jira/webhook/d64ebd66-c37b-11ee-95d9-da7ad0900005   ☐

Connect Jira to Monitor Notifications

Issues   ➕ New

🔍 Search issue templates

🌐

Create Jira issues from Datadog

Use @jira handles to create Jira issues from monitor alerts

New Issue Template ➕

Figure 10.32 – Datadog Case Management webhook configuration

14. Switch back to Jira and navigate to **System | WebHooks**.

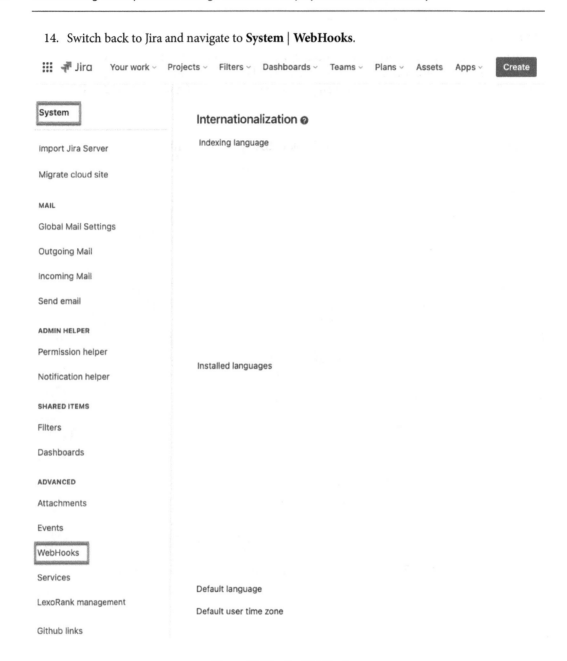

Figure 10.33 – Jira WebHooks

15. Select **Create a WebHook** and fill out the configuration parameters as follows:

    I.   Under **Name**, enter `Datadog Webhook`.

    II.   Set **Status** to **Enabled**.

    III.   For **URL**, enter the webhook URL copied to your clipboard from the Datadog Jira configuration step.

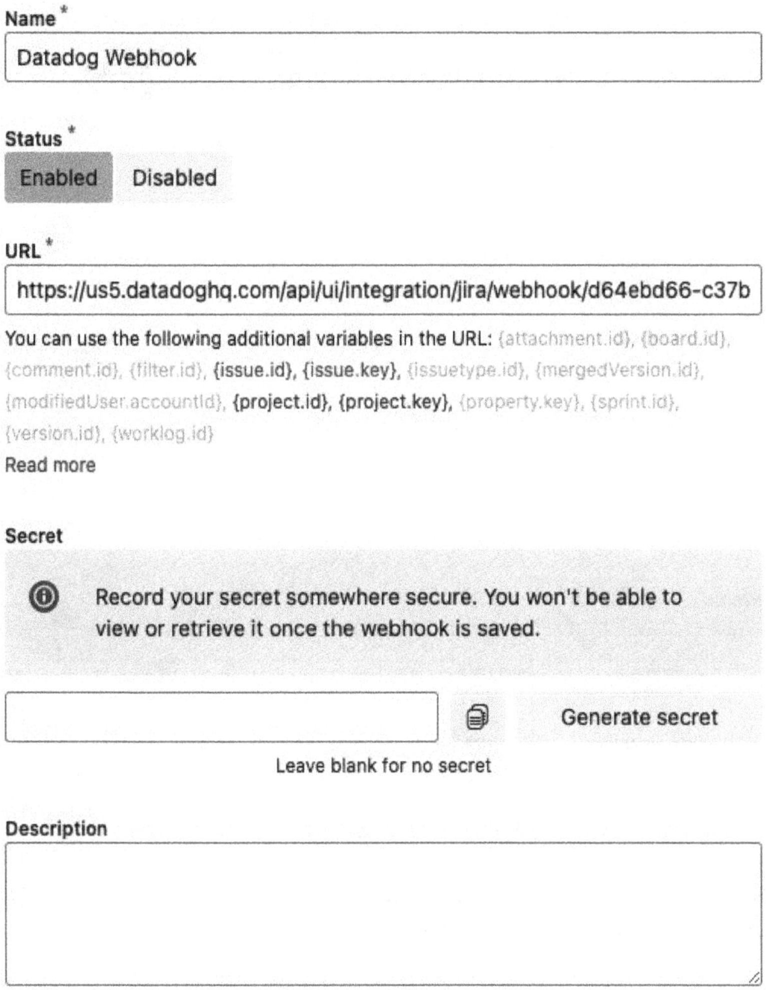

Name *

> Datadog Webhook

Status *

| Enabled | Disabled |

URL *

> https://us5.datadoghq.com/api/ui/integration/jira/webhook/d64ebd66-c37b

You can use the following additional variables in the URL: {attachment.id}, {board.id}, {comment.id}, {filter.id}, **{issue.id}**, **{issue.key}**, {issuetype.id}, {mergedVersion.id}, {modifiedUser.accountId}, **{project.id}**, **{project.key}**, {property.key}, {sprint.id}, {version.id}, {worklog.id}

Read more

Secret

    ⓘ   Record your secret somewhere secure. You won't be able to view or retrieve it once the webhook is saved.

| | 🗐 | Generate secret |

Leave blank for no secret

Description

Figure 10.34 – Jira WebHook configuration

    IV.   Under **Issue related events**, specify the project you want to associate with the Datadog webhook.

    V.   For **Issue**, check the **created**, **updated**, and **deleted** boxes.

      VI.   Under **Project related events**, check the **deleted** box.

      VII.   Click the **Save** button.

Issue related events

You can specify a JQL query to send only events triggered by matching issues. The JQL filter does not apply to events under the Issue link and Filter colu

> ✅ project in (ACR)

Syntax help

Issue	Worklog	Comment	Entity property	Attachment	Issue link	Filter
☑ created	☐ created	☐ created	☐ created or updated	☐ created	☐ created	☐ created
☑ updated	☐ updated	☐ updated	☐ deleted	☐ deleted	☐ deleted	☐ updated
☑ deleted	☐ deleted	☐ deleted				☐ deleted

User related events

User
☐ created
☐ deleted
☐ updated

Jira configuration related events

Features status change (enabled/disabled)
☐ voting
☐ watching
☐ unassigned issues
☐ subtasks
☐ issue links
☐ time tracking
☐ time tracking provider

Project related events

Events for projects and project versions.

Project	Version	Issue Type
☐ created	☐ released	☐ created
☐ updated	☐ unreleased	☐ updated
☑ deleted	☐ created	☐ deleted
☐ soft_deleted	☐ moved	

Figure 10.35 – Jira WebHook configuration (further details)

16. With the webhook created in Jira, switch back to your Datadog application. Navigate to the **Case Management** module and add a new project with an integration for Jira.

17. For the project Jira integration, ensure the following parameters are set:

      I.   Jira is enabled for this project.

      II.   Under **Jira (Atlassian) Account**, enter your Jira Cloud URL.

      III.   Set **Jira project name** to the Jira project to associate with this Datadog Case Management project.

IV.   Enter the issue type to create in Jira from this Case Management project.

V.   Enable **Automatically create and sync Jira issues**.

VI.   Configure your desired sync options for the bug fields.

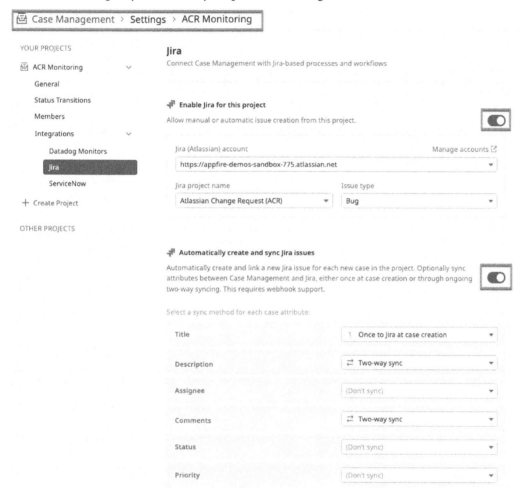

Figure 10.36 – Datadog Case Management project Jira integration

Important note

Now we can test the Datadog integration. We should be able to add a new case incident to our Datadog project, which will automatically create an associated Jira issue.

18. Navigate to your Datadog Case Management project and select the **New Case** button.

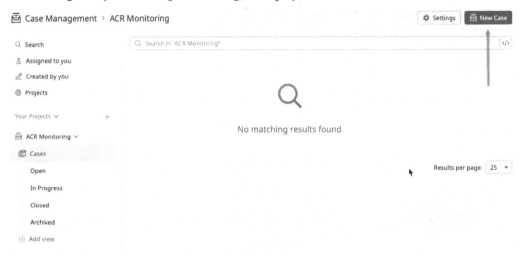

Figure 10.37 – Creating a new case

19. A new case pop-up window will be displayed. Fill out the **Title** field, add a description under **Description**, and finally click the **Create Case** button.

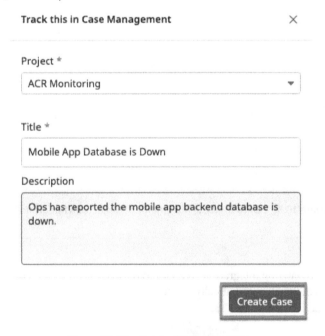

Figure 10.38 – Create Case parameters

A new case will be created in Datadog with an **OPEN** status. We can also see a **Jira Issue** button available. This indicates that a Jira issue has been created and linked back to this Datadog case. Click on the **Jira Issue** button to see the associated issue in Jira.

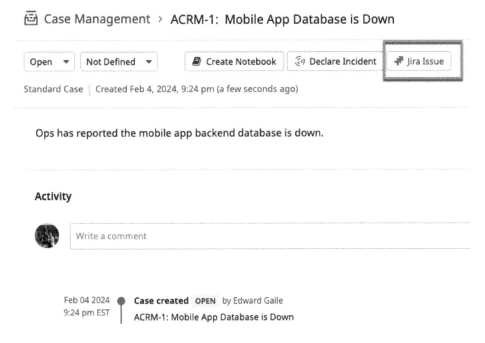

Figure 10.39 – New case displayed with Jira Issue button

We can see the Jira issue created with the associated Datadog case in the **Web links** section.

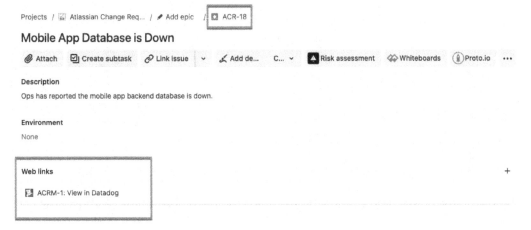

Figure 10.40 – Jira issue created by Datadog

We have now successfully integrated Datadog alert monitoring with our Jira application. Alerts can now automatically create Jira issues and add valuable insights to the overall DevOps process.

11

Monitoring Component Activity and Metrics Through CheckOps in Compass

Compass allows you and your teams to understand how your software all works together. In a distributed software architecture (microservices), there are many pieces that all come together to build something of value for your customer. Keeping track of all those pieces, and understanding who owns them and how they integrate, can be a challenging task. Compass makes it easy for software teams to catalog their components and be able to visually monitor their components and pull useful metrics.

First, we'll set up Compass, a separate Atlassian product that integrates with the rest of the Atlassian DevOps tools. Once you have Compass available, you'll create components, manage your team, create views of your software components, and create metrics to help your team understand how their software components all come together.

This chapter has the following recipes:

- Configuring Compass
- Importing distributed architecture components using a CSV file
- Integrating Compass with Bitbucket Cloud
- Understanding configuration as code in Compass
- Creating a developer platform with Compass
- Measuring DevOps health with Compass
- Utilizing templates in Compass
- Implementing developer CheckOps in Compass

Technical requirements

You will need the following:

- Jira
- A Git repository with your team's components
- Bitbucket Cloud

Configuring Compass

Compass is a standalone Atlassian product that you will need to purchase on its own. Before you can configure Compass, you will want to make sure that your company has a valid subscription to the Compass product.

Getting ready

To begin with the recipe, we need to get the setup ready by using the following steps:

1. Go to (`https://www.atlassian.com/try/cloud/signup?bundle=compass`).

   Once there, you'll want to sign in using your Google email, work email, or any other email by clicking on **Sign up with email**.

> Tip
>
> If you have used any other Atlassian product such as Jira, Confluence, or Bitbucket, you'll want to use the same account/ID that you use to log in to those products. Ideally, you should be using the same Atlassian ID that you have been using in the previous chapters.

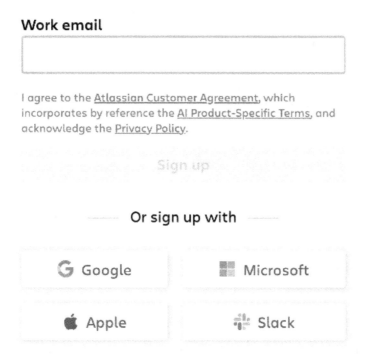

Figure 11.1 – Atlassian sign-in screen

2. Once you have signed in with your Atlassian ID, you'll select the site where you want to add Compass. You'll agree to the Atlassian Cloud Terms of Service and Privacy Policy.

> **Important note**
>
> Using Compass for all the examples in this book, you only need the **Free plan**. There is no credit card required and if you have three or fewer users, you will not have to pay for Compass. You will miss out on some of the features that are reserved for the paid tiers, but you will be able to follow along with the rest of the book with only the free version.

3.   Click on **Get started** to start configuring Compass.

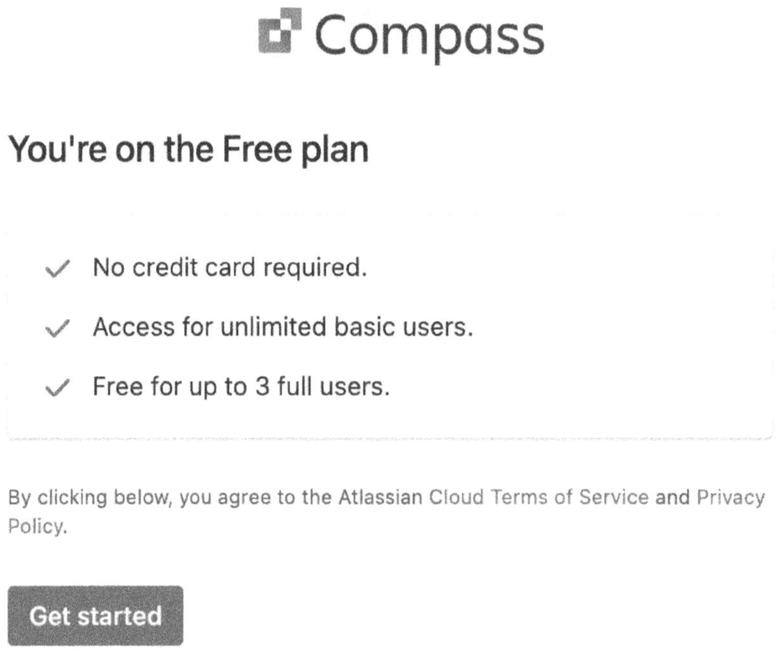

You're on the Free plan

✓   No credit card required.

✓   Access for unlimited basic users.

✓   Free for up to 3 full users.

By clicking below, you agree to the Atlassian Cloud Terms of Service and Privacy Policy.

Get started

Figure 11.2 – Compass Free plan

After a few moments, Atlassian's robots are going to deploy your very own instance of Compass and you'll be able to get started with creating your components and teams.

How to do it...

Before Compass can be a useful tool for your team, you need to populate it with components. These components are going to track all the various elements that make up your software stack. Components can be libraries, services, applications, dependencies, or anything that your product needs from a software perspective. Toward the end of this chapter, we have a recipe (*Creating a developer platform with Compass*) that covers how Jira and Compass connect to display Compass components within Jira. This is all automated by Atlassian and if your components exist within Compass, they will automatically be displayed in Jira.

There are two different ways to input components into Compass. You can manually create them, which is what we are going to do in this section, or you can import them. We are going to discuss importing components in the next section. For now, let's create some components the manual way.

Follow these instructions to configure the component catalog within Compass:

1. On the navigation bar across the top of Compass, click on the **Create** button. This is going to give you the option to create a variety of different elements in Compass. For this very first step, we want to select **Component**.

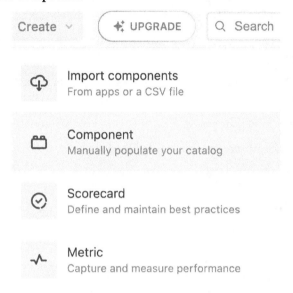

Figure 11.3 – Manually creating a new component

Important note

Creating a component can be a bit of an art. You have a variety of types to choose from that are built into Compass. You'll want to select the type that best describes the component you are trying to create. As discussed in the introduction to this section, a component is a piece of your software picture. Since software is complex, it is composed of many different components. You want to do your due diligence here and capture the components that appropriately describe your software.

---

2. Once you click on **Component**, you will be prompted to pick the type of component you want to create. Pick from any of the following available options per your choice:

    *   **Service** – An independently deployable software unit, usually operated by a person or team

    *   **Library** – A reusable collection of objects, functions, and methods

    *   **Application** – A fully-packaged application, such as a mobile application, desktop application, or a CLI tool

- **Capability** – A higher-level product functionality that an end user understands and sees value in

- **Cloud resource** – An entity or service provided by a cloud vendor, with consumer-managed configuration and monitoring

- **Data pipeline** – A sequence of tools and processes that gets data from a source to a target system

- **Machine learning model** – An algorithm that identifies patterns in a set of data and makes predictions over it

- **UI element** – Reusable building blocks of a design system that together create patterns and user experiences

- **Website** – One or more web pages under a domain, mainly consisting of read-only content

- **Other** – A generic software component described as *Other*

The following screenshot shows some of the listed options:

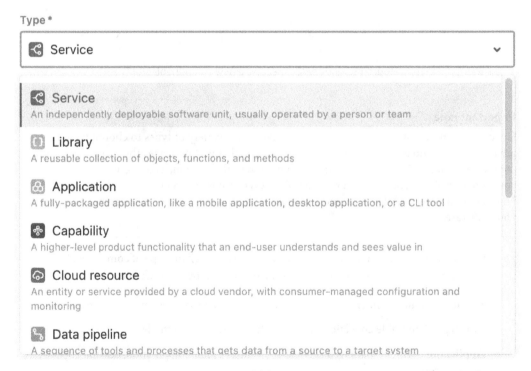

Figure 11.4 – Available component types

Important note

As you can see, there is an option for any part of your software. Select the type that most appropriately identifies with the component you are trying to capture. When first starting out, this may be challenging to do, because you might not have all the information captured. This is a good problem to have because part of having a healthy software product requires the team to have a 100% mapping of how all their software comes together. Having gaps in capturing your components can create blindsides that may have detrimental impacts later on in a software product's life cycle.

3. Once you have picked the component type, it is time to name it something.

Figure 11.5 – Naming the new component

Important note

After you have a suitable name, it is time to give someone ownership of the component. One of the most important things about going through this activity is that every component should be owned by someone or some team. This isn't a required field, but it's a field that is highly recommended to have populated. Having a team be responsible for a component will give someone ownership of that component. They'll be responsible for knowing their component inside and out. When bugs start being discovered, you'll want to know who owns the component. If a component isn't owned by anyone, that's okay, you can skip this part, but it is highly recommended that you start assigning owners to your components.

4. Select the **Owner team** for this component.

Figure 11.6 – Owner of the new component

Since this is our first component, we may not have any teams just yet, so for now, you can skip this part. After we create the teams in a future section, we'll be able to edit the existing components and assign them to the appropriate team. Alternatively, you can skip down to the *There's more…* section and create a team before proceeding.

> **Important note**
>
> If you or someone else has created a team in Jira or Confluence, those teams will be visible in Compass and will be selectable here.

5. The final piece of information you can provide is a link to the repository.

> The source code or configuration for this component. Add a repository and install a source code app to populate events and metrics. Learn more about getting metrics from an integrated tool

Figure 11.7 – Information about adding source code for a component

> **Tip**
>
> This step is optional but it's recommended that you provide a link to get the full effect and power of Compass.

6. Once you have finished putting in the information for your new component, it is time to click on the **Create** button:

Figure 11.8 – Confirmation button to create the new component

7. If you wish to create another component at this time, you can also click on the **Create another component** checkbox. This will retain the component type and **Owner team**, but the **Name** and **Repository** fields will be cleared out for your next component.

After you have created your first component, feel free to add the rest of your components. If you want to view all the components that you have created so far, simply click on the **Components** button in the navigation bar and you'll be redirected to the **Components** page.

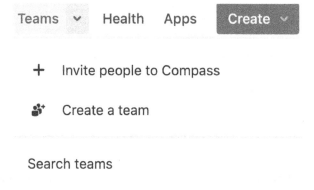

Components

Create **Import**

Search and discover software components across your distributed architecture. Learn more about components

🔍 Search component information

Type ∨   Tier ∨   Owner team ∨   Label ∨   Scorecard ∨   Lifecycle stage ∨   More filters +

1 component

Component	Description	Tier	Owner team	Scorecards
microservice name		TIER 4	Developer Team A	2 FAILING

Figure 11.9 – Confirmation of the newly created component

You have officially created your very first component in Compass! In the next section, we will create a team.

There's more...

Once you have your first component created, a team is needed to continue configuring Compass for the first time. In the previous section, you may have skipped adding a team to your component if you didn't already have a team. Let's look at how to create your first team in Compass:

1. On the navigation bar across the top of Compass, click on the **Teams** button. If you click on the down arrow, this will prompt you to invite people to Compass or give you the option to create a team; you can also search for any existing teams.

Teams ∨   Health   Apps   **Create** ∨

+ Invite people to Compass

👥 Create a team

Search teams

Figure 11.10 – Team portal within Compass

> **Important note**
>
> Inviting people to Compass requires a license. If you click on that button, an email is going to be sent to the recipient, but until your Atlassian administrator grants a license to that individual, they will not be able to access Compass. If you are the Atlassian administrator, you'll be able to grant a license, but most likely, you are not an Atlassian administrator, so this operation will not take effect until your Atlassian administrator adds the user themself.

2. Assuming your team members already have Compass licenses, click on **Create a team**. This is going to display a window where you'll be able to create a team and add members to it.

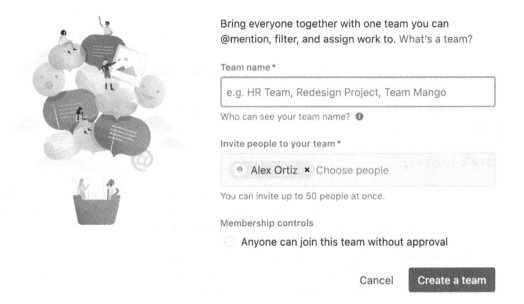

Figure 11.11 – Creating a new team

3. The first thing you want to do here is give your team a name. This can be anything you want it to be, but it should be something descriptive that will be obvious to anyone else looking at the team names.

    Once you have the team name, you need to add people. You can add up to 50 people at a time.

4. Next, you can grant others the ability to join your team without your approval. This would not be a good practice and it is entirely up to you whether you leave this option enabled or not.

5. Finally, click on the **Create a team** button, or hit **Cancel** if you no longer want to create a team.

6. After your team has been created, it will be visible in the **Teams** section of Compass. You can get to the **Teams** section by clicking on the **Teams** button from the navigation bar.

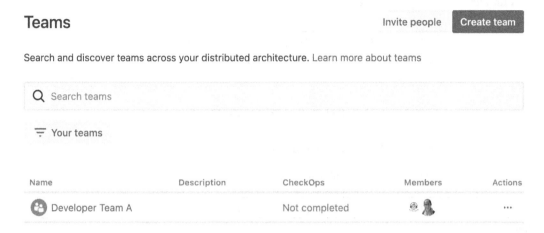

Figure 11.12 – Reviewing the newly created team

From the **Teams** page, you can also invite people and create additional teams.

> **Important note**
>
> The team you just created isn't just for Compass. This team is available in other Atlassian products such as Jira and Confluence. Similarly, if you or someone else created a team in Jira or Confluence, then you should expect to see those teams in Compass as well.

Importing distributed architecture components using a CSV file

This recipe covers how to quickly import many components all at the same time by using a CSV file. This is much better than manually creating components as most teams will have dozens, if not hundreds of components that need to be tracked in Compass.

Getting ready

Manually importing every component for your software is going to take a very long time. Fortunately, there is a way to import your components using a CSV file. At a minimum, your CSV file will require two important columns of information.

The required data is as follows:

Column name	Required input
Name	Any name you want
type	`SERVICE, LIBRARY, APPLICATION, CAPABILITY, CLOUD_RESOURCE, DATA_PIPELINE, MACHINE_LEARNING_MODEL, UI_ELEMENT, WEBSITE, OTHER`

Table 11.1 – Required data

There are many other fields/columns that you can include with your import, but they aren't required.

Here's a list showing what those optional field/column names include:

Life cycle stage	`Pre-release, Active, Deprecated`
tier	`1, 2, 3, 4`
description	Description of what the component is
labels	Any additional helpful information to categorize the component
owner team	Any team you want, following special guidance covered in the next section
repository	The repository where the component is
chat channels	Information on any chat channels that may involve the component
projects	The Jira project that references/utilizes the component
documentation	Confluence documentation for the component
dashboards	Links to any dashboards that reference the component
on-call schedules	Links to any on-call schedules for the team responsible for the component
other links	Anything you want
Custom Fields	Anything you want, but needs to start with `custom:`

Table 11.2 – Optional CSV fields/columns

The most important thing for your CSV file is to have the header row correct. Each value in the header row should correspond to one of the aforementioned fields. Keep in mind that all the values in the header row should be lowercase, such as in the following figure:

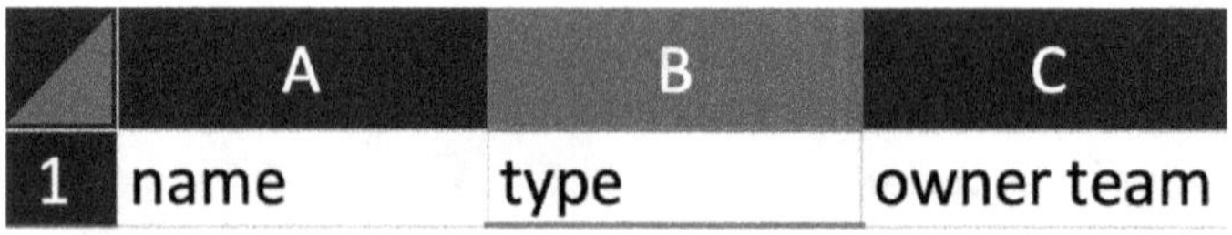

Figure 11.13 – Header column example

With the prerequisites known to us, let's get to the practical part.

How to do it...

Since we will assign the components to an owner team, we'll need to get the team's ID before proceeding. We will use the following steps:

1. In Compass, click on the **Teams** button and select the team you want.

2. Once the team has been selected, you'll be redirected to the team's page where you'll be able to click on **...** to obtain the owner ID. Click on **Copy owner ID**.

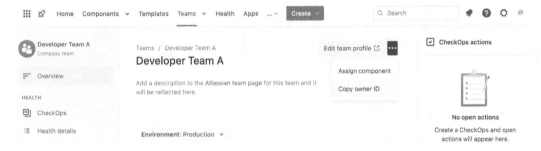

Figure 11.14 – Selecting the team to obtain the owner ID

3. With the owner ID in hand, it's time to start populating our CSV file.

	A	B	C
1	name	type	owner team
2	second microservice	SERVICE	ari:cloud:teams::team/1e2e09c8-15ba-4a43-99c8-33090d12c760
3	third microservice	SERVICE	ari:cloud:teams::team/1e2e09c8-15ba-4a43-99c8-33090d12c760
4	application 1	APPLICATION	ari:cloud:teams::team/1e2e09c8-15ba-4a43-99c8-33090d12c760
5	Website 1	WEBSITE	ari:cloud:teams::team/1e2e09c8-15ba-4a43-99c8-33090d12c760

Figure 11.15 – Populating the CSV file with the owner ID

4. Once you are done capturing all your components, save your file as a CSV. Keep in mind that you can only import up to 500 components in each file. If you have more than that, you'll have to split your CSV into multiple files.

Figure 11.16 – Saving the CSV file

5.   In Compass, click on the **Create** button and select **Import components**.

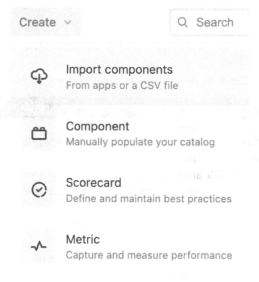

Figure 11.17 – Importing the components

6.   When the next window appears, scroll all the way to the bottom and click on **Import** on the right of the **CSV file** option.

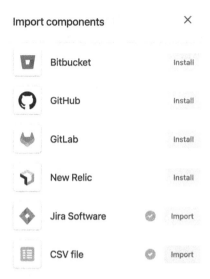

Figure 11.18 – Importing the CSV file

7. Upload your file by dragging it to the **Upload** section.

Figure 11.19 – Uploading the CSV file

8.   If your CSV file is not in the correct format, the review will not pass. In that case, I recommend you download the sample CSV file from the previous screen. If you get all green checkmarks, then you can proceed to the next step by clicking on the blue **Review** button.

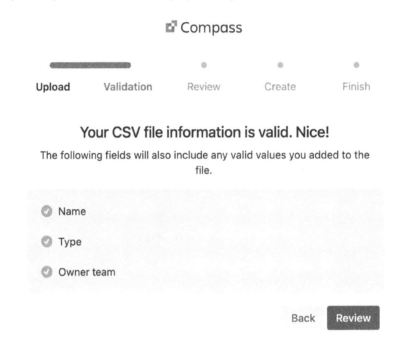

Figure 11.20 – CSV file validation

9. The **Review** step will analyze your file and determine how many components are going to be imported. Click on the blue **Create** button to proceed.

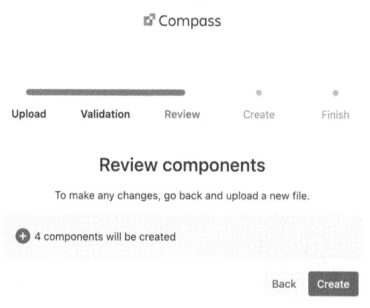

Figure 11.21 – Reviewing the components

10. Compass will take a few seconds to import your components and then you'll be able to view your newly created components after the import is complete.

Click on **View components** and you'll be able to see all your components and verify that the information was imported correctly.

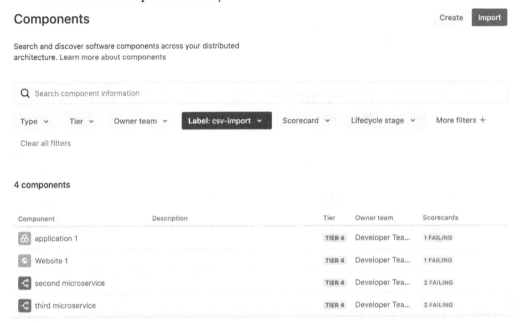

Figure 11.22 – Verifying the components in Compass

Now that your components have all been imported into Compass, it is time to learn how to create even more dynamic components.

Integrating Compass with Bitbucket Cloud

One of the advantages of using Compass is that it connects with Bitbucket Cloud. This connection will allow Compass to communicate with Bitbucket and utilize Bitbucket as a single source of truth for your component data. You will need to have an existing Bitbucket repository, which will be used to integrate with Compass.

How to do it...

In Compass, you will establish a connection between your Git repository and Compass using the following steps:

1. In Compass, click on **Apps**, located on the navigation bar. This will open a screen where all the apps that you can integrate with Compass are available. Select Bitbucket by clicking **Install** within the **Bitbucket** tile.

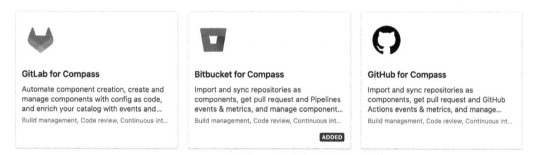

GitLab for Compass

Automate component creation, create and manage components with config as code, and enrich your catalog with events and...

Build management, Code review, Continuous int...

Bitbucket for Compass

Import and sync repositories as components, get pull request and Pipelines events & metrics, and manage component...

Build management, Code review, Continuous int...   **ADDED**

GitHub for Compass

Import and sync repositories as components, get pull request and GitHub Actions events & metrics, and manage...

Build management, Code review, Continuous int...

Figure 11.23 – Selecting the Bitbucket tile

2. The installation should only take a few seconds. After it has been installed, the Bitbucket tile will now allow you to configure (or uninstall). Click on the **Configure** button.

Bitbucket

By Atlassian

Connect Bitbucket to Compass to import components from Bitbucket, manage components with config-as-code, use templates, gain visibility into deployments, and get metric data. Learn more about integrating with Bitbucket Cloud

Configure   Uninstall

Figure 11.24 – Clicking on Configure

3. On the configuration screen, click on **Connect to Bitbucket**.

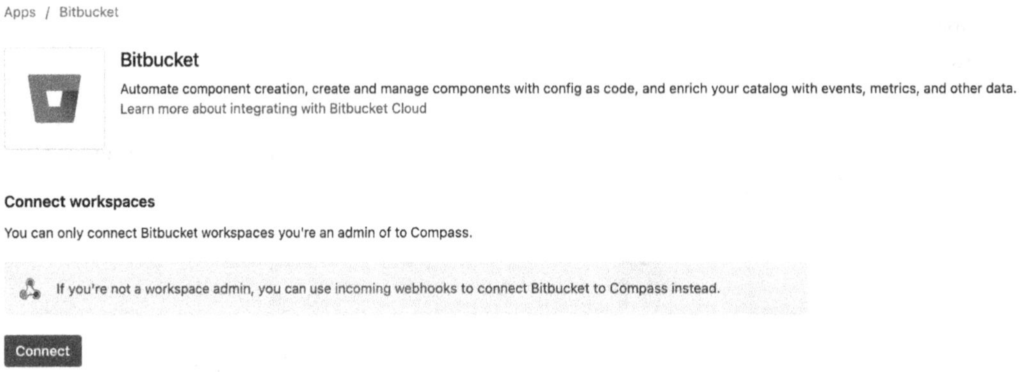

Apps / Bitbucket

Bitbucket
Automate component creation, create and manage components with config as code, and enrich your catalog with events, metrics, and other data.
Learn more about integrating with Bitbucket Cloud

Connect workspaces

You can only connect Bitbucket workspaces you're an admin of to Compass.

 If you're not a workspace admin, you can use incoming webhooks to connect Bitbucket to Compass instead.

Connect

Figure 11.25 – Connecting to Bitbucket

4. Click on the yellow **Continue** button when the popup appears.

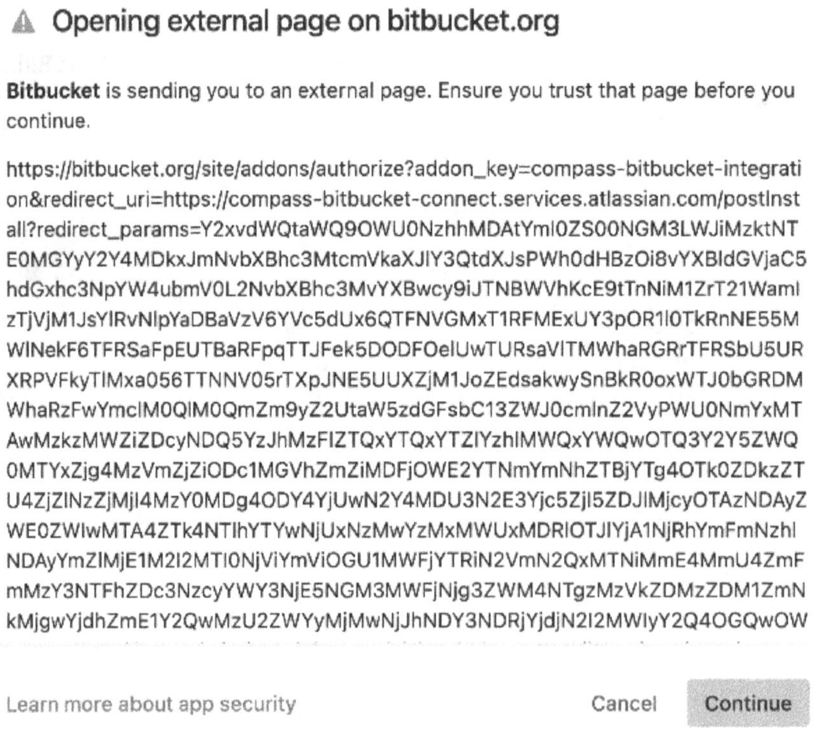

⚠ **Opening external page on bitbucket.org**

Bitbucket is sending you to an external page. Ensure you trust that page before you continue.

https://bitbucket.org/site/addons/authorize?addon_key=compass-bitbucket-integrati
on&redirect_uri=https://compass-bitbucket-connect.services.atlassian.com/postInst
all?redirect_params=Y2xvdWQtaWQ9OWU0NzhhMDAtYml0ZS00NGM3LWJiMzktNT
E0MGYyY2Y4MDkxJmNvbXBhc3McmVkaXJlY3QtdXJsPWh0dHBzOi8vYXBldGVzdC5
hdGxhc3NpYW4ubmV0L2NvbXBhc3MvYXBwcy9iJTNBWVhkcE9tTnNiM1ZrTW21Wm
zTjVjM1JsYlRvNlpYaDBaaVZ6YVc5dUx6QTFNVGMxT1RFMExUUY3pOR1l0TkRnNE55M
WlNekF6TFRSaFpFUTBaRFpqTTJJFek5DODFOelUwTURsaVITMWhaRGRyTFRSbU5UR
XRPVFkyTIMxa056TTNNV05rTXpJNE5UUXZjM1JoZEdsakwySnBkR0oxWTJ0bGGRDM
WhaRzFwYmclM0QlM0QmZm9yZ2UtaW5zdGFsbC13ZWJ0cmlnZ2VyPWU0NmYxMT
AwMzkzMWZiZDcyNDQ5YzJhMzFIZTQxYTQxYTZIYzhIMWQxYWQwOTQ3Y2Y5ZWQ
0MTYxZjg4MzVmZjZiODc1MGVhZmZiMDFjOWE2YTNmYmNhZTBjYTg4OTk0ZDkzZT
U4ZjZINzZjMjl4MzY0MDg4ODY4YjUwN2Y4MDU3N2E3Yjc5ZjI5ZDJIMjcyOTAzNDAyZ
WE0ZWIwMTA4ZTk4NTIhYTYwNjUxNzMwYzMxMWUxMDRIOTJlYjA1NjRhYmFmNzhl
NDAyYmZIMjE1M2I2MTI0ONjViYmViOGU5MWWFjYTRiN2VmN2QxMTNiMmE8MmU4ZmF
mMzY3NTFhZDc3NzcyYWY3NjE5NGM3MWFjNjg3ZWM4NTgzMzZVkZDMzZDM1ZmN
kMjgwYjdhZmE1Y2QwMzU2ZWYyMjMwNjJhNDY3NDRjYjdjN2I2MWIyY2Q4OGQwOW

Learn more about app security   Cancel   Continue

Figure 11.26 – Bitbucket authorization

5. On the following screen, select the Bitbucket workspace you want to connect with and then click on **Grant access**.

Compass requests access

This app is hosted at https://compass-bitbucket-
connect.services.atlassian.com

Read and modify your account information

Access your repositories' build pipelines and configure their
variables

Read and modify your workspace's project settings, and read
and transfer repositories within your workspace's projects

Read and modify your repositories and their pull requests

Access and edit your workspaces/repositories' runners

Read and modify your repositories' webhooks

Authorize for workspace

DevOps Book (devops-book)   ⌄

Allow Compass to do this?

Grant access   Cancel

Figure 11.27 – Granting Bitbucket access to Compass

6. Once your workspace has been connected, you then need to select the repository that will contain your component data. Click on the blue **Select repositories** button.

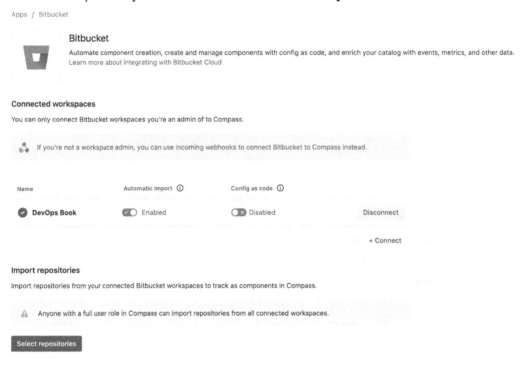

Figure 11.28 – Selecting the Bitbucket repository

Compass will show you the repositories available in your workspace and you just need to select them. Choose the component type in the repository and then click on the blue **Select** button.

Figure 11.29 – Bitbucket repository selection

7. Select the checkbox for **Set up configuration files for all repositories during import**.

☑ Set up configuration files for all repositories during import

Figure 11.30 – Setting up configuration files during import

This is going to add the `compass.yml` file to each repository, which is critical for Compass to utilize Bitbucket as the single source of truth for your component metadata. Once you have made your selections, you will see the following screen, which will then prompt you to start the import process:

Apps / Bitbucket

Bitbucket   Uninstall
Connect Bitbucket to Compass to import components from Bitbucket, manage components with config-as-code, gain visibility into deployments, and get metric data.
Learn more about integrating with Bitbucket Cloud

Set up configuration files

Manage components with configuration files in the connected Bitbucket workspace and sync any updates back to Compass.
Learn more about managing components with config-as-code

You can choose to let Compass set up the configuration files for all your components during the import. Or, you can import your components first, and then set up the configuration files manually later.

☑ Set up configuration files for all repositories during import

ⓘ When you start the import, Compass adds the compass.yml configuration file to each repository and raises pull requests. Approve and merge the pull requests for the repositories to sync with Compass.

Repository	Repository description	Component type
snakechart		Service ∨

Edit selection   Start import

Figure 11.31 – Starting the Bitbucket import process

8. Click on the blue **Start import** button and you will be notified with **Import completed**.

9. Click on **Done** when you are finished connecting Bitbucket with Compass.

You have successfully integrated Bitbucket Cloud with Compass.

Understanding configuration as code in Compass

In the previous two recipes (*Configuring Compass* and *Importing distributed architecture components using a CSV file*), you learned how to manually create and import components. Both methods require a lot of user input and can be prone to errors. Not only that, but if your component data changes in your repository, someone needs to go into Compass and update the component data there. A better way would be to have the code in your repository update the component information. This can be achieved by utilizing **configuration as code (CaC)**.

Once you connect your Bitbucket repository with Compass, you'll be able to use a special YAML file for each of your components. This YAML file exists for each component within your code repository and, best of all, it is version-controlled. Instead of having to update Compass manually or via CSV files, you'll simply be able to manage all your component data within your code repository and Compass will take care of the rest.

How to do it...

In your code repository, identify the `compass.yml` file. You'll be providing your component metadata here, which will then make this component(s) a managed component within Compass, which basically means that the data will come from the `.yml` file and not manually from Compass.

The following example will be for a single component that is tracked in your Bitbucket repository using the `compass.yml` file for that component. Each component will need its own `compass.yml` file. Make sure that each unique `compass.yml` file for each unique component is in its own subfolder within your code repository. Let's use the following steps for it:

1.   Within your code repository, find or create the `compass.yml` file. The following is the sample `compass.yml` file that is created when Bitbucket is connected with Compass. If you have not completed this connection, skip down the *Integrating Compass with Bitbucket Cloud* section.

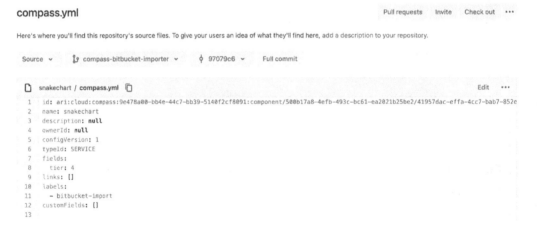

Figure 11.32 – compass.yml example

> **Important note**
>
> You still need to create the component in Compass. Once the component is created, you will be able to retrieve the unique ID for the component, which will then be used by the `compass.yml` file to manage the component's metadata. Bitbucket will create a component for you when you connect the repository to Compass, but if you want to track/add multiple components, you are either going to need multiple repositories or you'll want to manually create the components in Compass and then reference them in multiple `compass.yml` files, sprinkled as appropriate in your repository.

In the following points, we are first going to break down and explain the contents of this file and then we are going to commit it to Bitbucket, which will then push the data to Compass as a managed component.

A. `id` (required): This is the unique identifier that Compass uses to track your component. Obtain the ID by clicking into the component within the component's catalog in Compass and clicking on **Copy component ID**.

Figure 11.33 – Obtaining the component ID

B. `name` (required): This is going to be the name of your component. This should match what you have in Compass.

C. `description` (optional): This is the description of what your component does/is. If you want to write multiple lines to describe your component, you need to start the description with the text `| -`.

D. `ownerId` (optional): If you want to assign the component to a team, you will need to retrieve the value from Compass, find the team that should own the component, and click on **Copy owner ID**

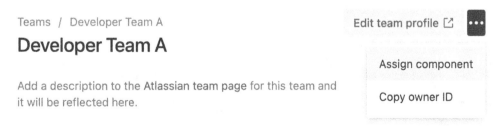

Figure 11.34 – Assigning the component to a team

E.   `configVersion` (optional): Leave this as `1`, as that is the only value you can provide.

F.   `typeID` (required): This needs to match with the available component types:

- APPLICATION

- SERVICE

- CAPABILITY

- CLOUD_RESOURCE

- DATA_PIPELINE

- LIBRARY

- MACHINE_LEARNING_MODEL

- OTHER

- UI_ELEMENT

- WEBSITE

The rest of the fields are completely optional and you can read Atlassian's documentation if you want to populate them (*E*).

2.   Once you have your file updated, all you need to do is commit your changes in Bitbucket by clicking on the **Commit** button.

Figure 11.35 – compass.yml commit example

You'll see the details of your changes.

Figure 11.36 – Commit example

3. Next, create a pull request to bring your changes back to the **master** or **develop** branch.

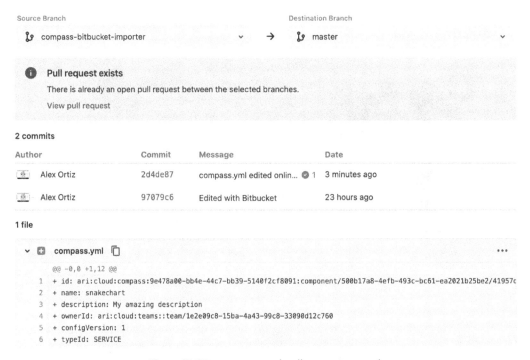

Figure 11.37 – compass.yml pull request example

4. Approve and merge the pull request using your typical pull request process.

Back in Compass, for your managed component, you'll see the updated information.

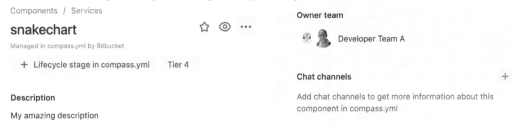

Figure 11.38 – Component confirmation in Compass

5. Finally, if you created your component in the Compass UI and you wish to minimize the level of effort it takes to create your compass.yml file, go to the component within the Compass UI and select **Config-as-code** from the left-hand side of the selected component.

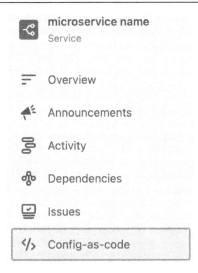

microservice name
Service

Overview

Announcements

Activity

Dependencies

Issues

</> Config-as-code

Figure 11.39 – compass.yml pull request example

You'll be able to download the `compass.yml` file that you can then utilize in your Bitbucket repository.

Components / Services / microservice name

Config-as-code

Manage and update components from a YAML configuration file alongside your code. Learn more about using YAML files to manage components.

1. Create compass.yml

compass.yml	Copy Download

```
 1  name: microservice name
 2  id: 'ari:cloud:compass:9e478a00-bb4e-44c7-bb39-5140f2cf8091:component/500b17a8-4efb-493c-bc61-ea2021b25be2/b8f165b5
 3  description: null
 4  configVersion: 1
 5  typeId: SERVICE
 6  ownerId: 'ari:cloud:teams::team/1e2e09c8-15ba-4a43-99c8-33090d12c760'
 7  fields:
 8    tier: 4
 9  links: []
10  relationships:
11    DEPENDS_ON:
12      - 'ari:cloud:compass:9e478a00-bb4e-44c7-bb39-5140f2cf8091:component/500b17a8-4efb-493c-bc61-ea2021b25be2/dedc27
13  labels: null
14  customFields: null
15
16  # Learn more about formatting compass.yml:
17  # https://go.atlassian.com/compass-yml-format
```

Figure 11.40 – Config-as-code example

This is ultimately the best way to manage and track your components within Compass. Since Compass components should represent the subsystems within your code, it makes sense to embed this `.yaml` file into your code base.

Now that all your components are in Compass, it is time to start leveraging Compass to ensure your components are up-to-date and in good working condition.

Creating a developer platform with Compass

Compass is designed to help you gain an entirely new appreciation of how all your code is connected. With all the components in Compass, it is time to build a developer platform. This platform is going to give you a unified view of all your components, libraries, services, applications, and documentation, and the health and status of whatever you track within Compass.

How to do it...

The first thing we want to do is connect components that are related and highlight their dependencies. These dependencies will show what other components a specific component relies on. Having this mapped out in Compass is crucial to understanding how your software architecture works. Understanding these relationships enables your team to have a better understanding of how components rely on each other to make your software work.

Let's use the following steps to create these dependencies:

1. Click on **Components** in the Compass header and select the component that you want to add a dependency to.

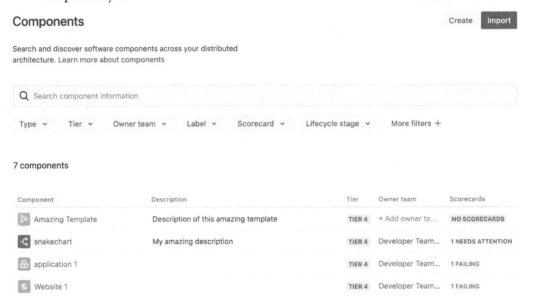

Figure 11.41 – Selecting from an existing component

2. On the left-hand side, click on **Dependencies**.

Figure 11.42 – Selected component details

3. In the drop-down box/field for **Depends on**, select the component that your selected components depend on.

Map your software infrastructure

Upstream and downstream dependencies of **microservice name** will appear here.

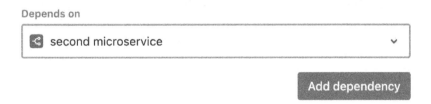

Figure 11.43 – Selecting component dependency

Click on **Add dependency**, and you will see the following screen:

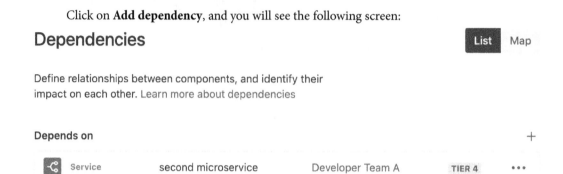

Figure 11.44 – Adding component dependency

From there, add as many dependencies as you need to appropriately map out how your components are connected to each other.

> **Important note**
>
> Compass allows you to view your dependencies in a list or map view. This will allow you to visualize any circular dependencies that may exist within your components. Also, please keep in mind that Compass allows a component to have a maximum of 25 dependencies. A maximum of 100 other components can depend on a single component.

There's more...

Compass has a built-in functionality that allows you to make announcements about any component in your Compass catalog. This is extremely helpful as teams can stay up-to-date about what is going on with the components that are important to them.

Compass includes a feature that lets you create announcements for any component in your Compass catalog. This is useful for keeping teams informed about important component updates.

To share relevant information with your team about a specific component, follow these steps to create an announcement in Compass:

1. Select a component and then click on **Announcements**.

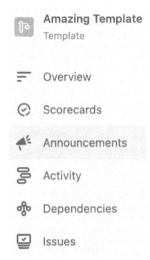

Amazing Template
Template

Overview

Scorecards

Announcements

Activity

Dependencies

Issues

Figure 11.45 – Making an announcement selection

2. Click on **Create announcement**.

Announcements

Share upcoming news or changes, and stay up to date with announcements that may affect this component.

From this component   From dependencies

Broadcast upcoming changes

Create announcements to share updates with any components that
depend on **microservice name**. For example:

- New features
- API deprecation
- Scheduled downtime

Create announcement

Figure 11.46 – Creating an announcement

3. Provide a name and target date for your announcement. Optionally, you can also provide a description. Then, click **Create**.

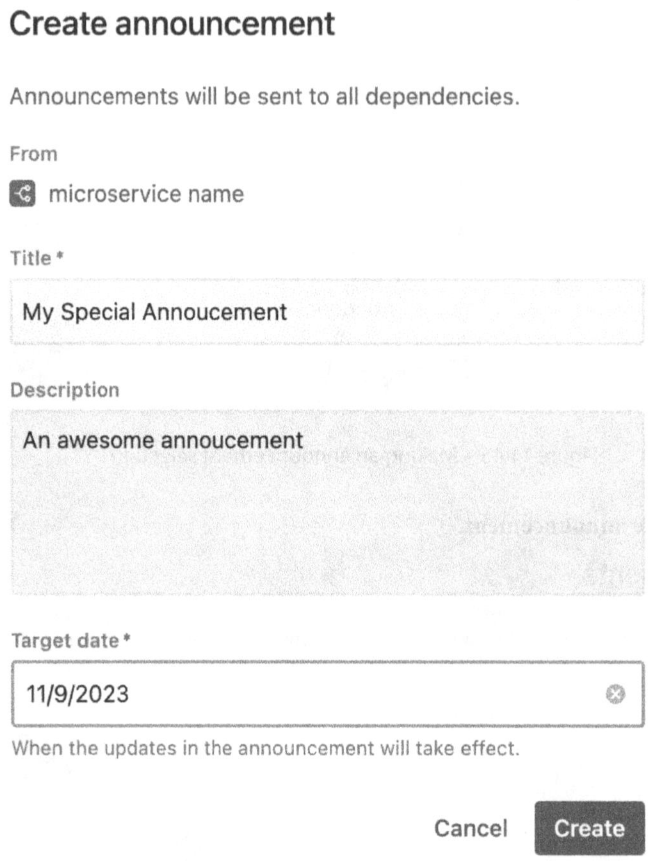

Create announcement

Announcements will be sent to all dependencies.

From

⟨⟩ microservice name

Title *

My Special Annoucement

Description

An awesome annoucement

Target date *

11/9/2023

When the updates in the announcement will take effect.

Cancel   **Create**

Figure 11.47 – Announcement creation example

Your announcement will now be viewable for this component. Any components that depend on this specific component will also be able to see the announcement. Also, team members will receive an email notification and see a notification badge within the Compass UI indicating that an announcement is available.

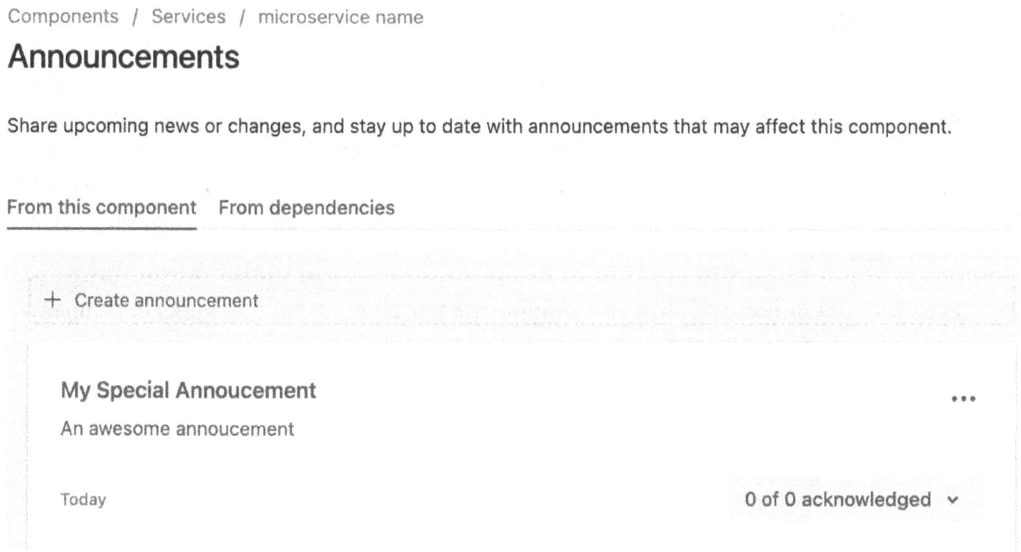

Components / Services / microservice name

Announcements

Share upcoming news or changes, and stay up to date with announcements that may affect this component.

From this component   From dependencies

+ Create announcement

My Special Annoucement

An awesome annoucement

Today   0 of 0 acknowledged ⌄

Figure 11.48 – Announcement review example

See also

This recipe has a few more elements that you should consider.

Monitoring component activity

The **Activity** page will show you what is going on with a component. In the Activity page, you can visualize deployments, incidents, flags, and any alerts that are associated with the selected component. This view will give you a real-time understanding of what condition a component is in. If the components are connected together via their dependencies, you can see that information as well.

Figure 11.49 – Component activity monitoring

Connecting with Jira

Jira can also connect with Compass, and you can visualize Compass data within Jira and vice versa. There used to be two different ways to connect Compass components with Jira. One was introduced in 2022. Users wanting to integrate Compass with Jira needed to enable a custom field within Jira that would pull in the components from Compass. That field and method were deprecated on May 31, 2024. As a result, the following steps are the best way to ensure that your Compass components can be used within Jira:

1. Go to your Jira project and click on **Components** (please note that this works in company-managed projects only).

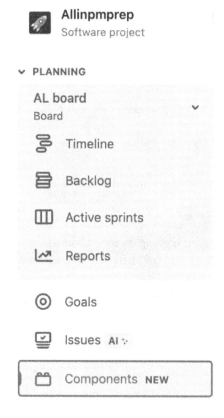

Figure 11.50 – Selecting components from a company-managed Jira project

2. Once in the **Components** section, you'll see that Jira is automatically connected with Compass. Click on **Go to issues**.

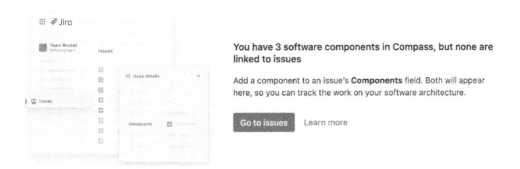

You have 3 software components in Compass, but none are linked to issues

Add a component to an issue's **Components** field. Both will appear here, so you can track the work on your software architecture.

Go to issues   Learn more

Figure 11.51 – Components menu when Jira and Compass are connected

3. Pick any issue and click on the **Components** field.

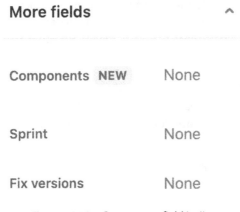

Figure 11.52 – Component field in Jira

4.   In the drop-down box/field, you will now see the components that exist in Compass.

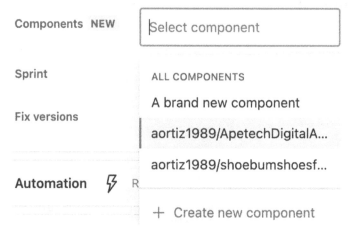

Figure 11.53 – Compass components in Jira

Connecting Jira with Compass allows your teams to be much more efficient and the best part is that you do not need to duplicate data. Instead of having to recreate all your components again in Jira, your team can leverage the same Compass components within Jira. If you use the Jira-only components, you will need to recreate those components in Compass. This will allow Jira to have the latest and most up-to-date components that are pulled in from Compass.

Measuring DevOps health with Compass

Connecting your components with Compass and mapping out their dependencies is great, but even all of that does not give you a full picture of the health and status of your components. With Compass, you can create scorecards to help visualize how components are doing.

How to do it...

Compass comes with a couple of scorecards out of the box. Those scorecards might be all you need, but in case you need additional scorecards, it is very easy to create them. Let's use the following steps to do that:

1.   In the navigation bar within Compass, click on **Health**.

Figure 11.54 – Health button to get to the scorecards

Once there, you'll be able to see the existing scorecards.

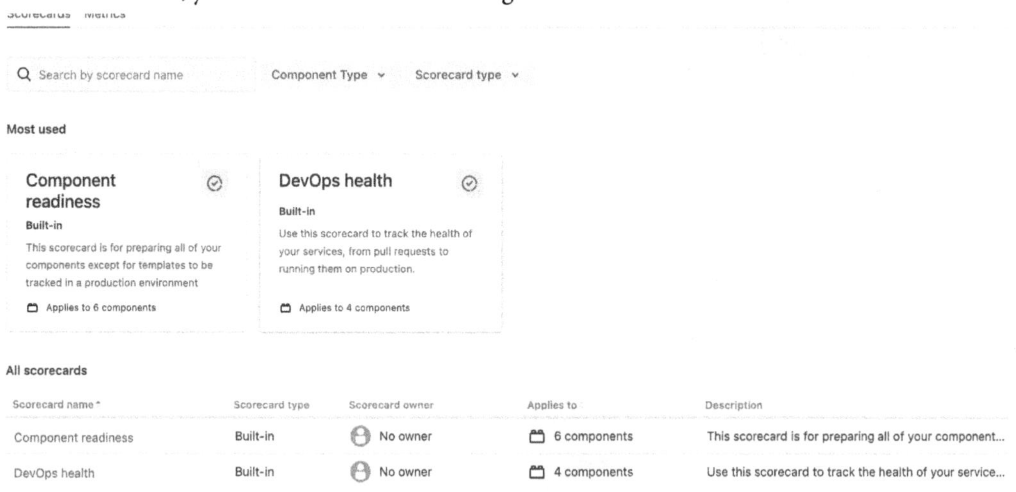

Figure 11.55 – Compass scorecards

2. Click on **Create scorecard** in the upper-right corner.

Figure 11.56 – Compass's menu option to create a scorecard

This will present you with a modal that will allow you to create your scorecard.

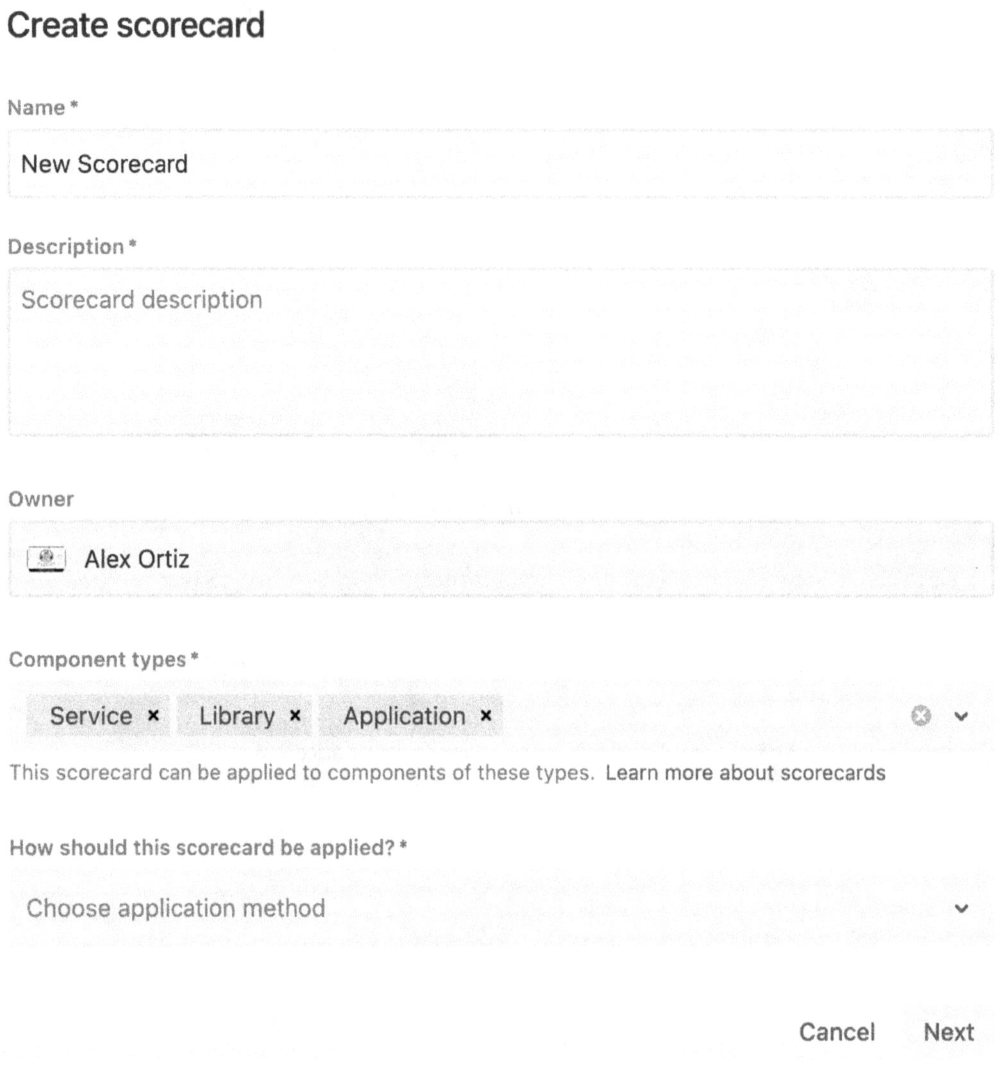

Create scorecard

Name *

New Scorecard

Description *

Scorecard description

Owner

Alex Ortiz

Component types *

Service ×   Library ×   Application ×

This scorecard can be applied to components of these types. Learn more about scorecards

How should this scorecard be applied? *

Choose application method

Cancel   Next

Figure 11.57 – Modal to create a scorecard

Provide a name, description, and owner, then select your component types and how the scorecard should be applied. Click **Next** to proceed.

3.   In the next screen, you will select metadata from the component to help you build out the criteria that will let your team determine how healthy the component is.

Create scorecard

Choose this scorecard's criteria. Your criteria percentages must total 100%. You can distribute the weight evenly or make some criteria weigh more than others. Learn how each criteria is used on a component.

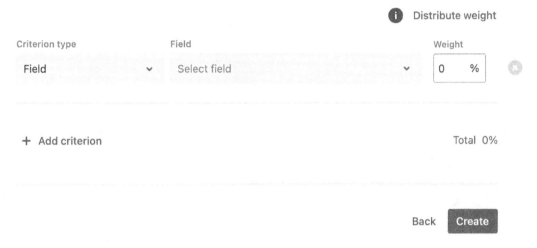

Figure 11.58 – Selecting metrics to be displayed in the scorecard

4.   As you build out your scorecard, you get to pick the **Criterion type**. Your options are **Field** or **Metric**.

Figure 11.59 – Selecting your criterion type

After that, you get to pick the **Field** or **Metric** that will be used as the criterion.

5. Finally, you pick your weight for your criterion.

> **Tip**
> Repeat steps 4 and 5 for all the criteria that you want to create for your scorecard. All your criteria must equal a 100% weight distribution. When you are done, your scorecard will look like the following screenshot:

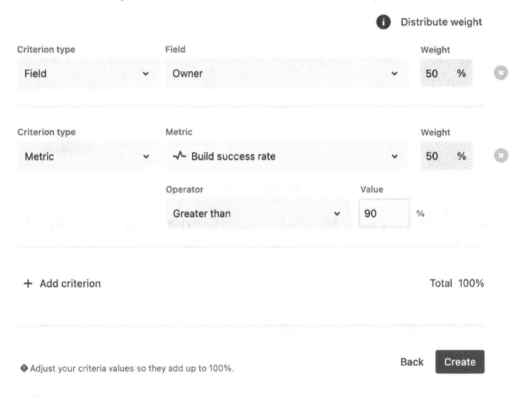

Create scorecard

Choose this scorecard's criteria. Your criteria percentages must total 100%. You can distribute the weight evenly or make some criteria weigh more than others. **Learn how each criteria is used on a component.**

ⓘ Distribute weight

Criterion type	Field	Weight
Field ⌄	Owner ⌄	50 % ⊗

Criterion type	Metric	Weight
Metric ⌄	⌁ Build success rate ⌄	50 % ⊗

	Operator	Value
	Greater than ⌄	90 %

\+ Add criterion   Total 100%

◆ Adjust your criteria values so they add up to 100%.   Back   **Create**

Figure 11.60 – Completing the criteria before creating a scorecard

6. Click **Create** when you are done.

Back in your component, you will be able to see that the scorecard has been applied to your component based on how you configured your scorecard.

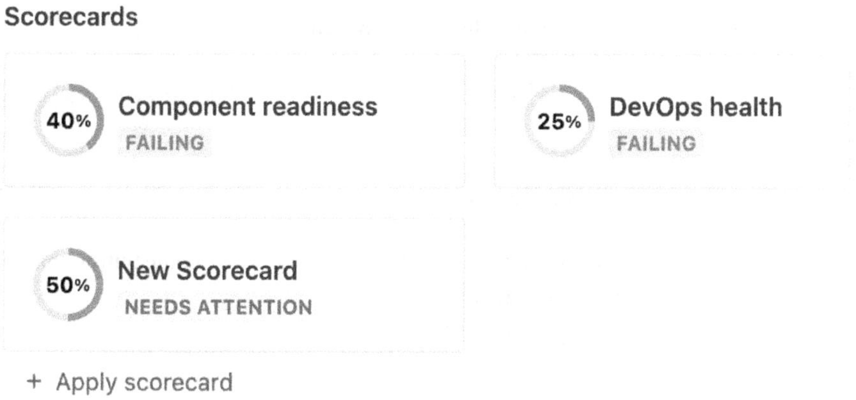

Scorecards

Figure 11.61 – Scorecard values in Compass

Experiment with all the different criteria available to you. Building these scorecards is what sets Compass apart from the rest of the Atlassian products. As your teams utilize the components in their code, the scorecards keep track of key metrics to help you all understand the health and status of your code.

Utilizing templates in Compass

Templates in Compass allow you to create boilerplate pieces of code that help streamline your software development process. These templates become components in Compass and can be reused as appropriate throughout your software stack. If your team is repeatedly having to create similar code or configurations, templates can really help your team be more efficient. Your team's code quality can also improve through the use of templates as you'll always have a solid foundation to get started with. These templates that you'll learn how to create don't have to be components; they can be any useful pieces of code that your team depends on frequently.

How to do it...

The first step in utilizing templates is to connect Compass with Bitbucket. Refer to the *Integrating Compass with Bitbucket Cloud* recipe if you have not already connected Bitbucket with Compass.

Once your repository is connected with Compass, it's time to create your first template. Let's do it using the following steps:

1. In the navigation bar within Compass, click on **Templates**.

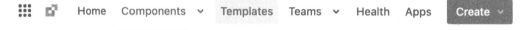

Figure 11.62 – Templates in Compass

Assuming this is the first time you create a template, you should see a blue button that prompts you to create your first template. Click on it.

No templates

Templates that you build and create components from
will appear here.

Figure 11.63 – Creating templates in Compass

2. Fill out the information, which looks exactly like when you are creating a new component manually.

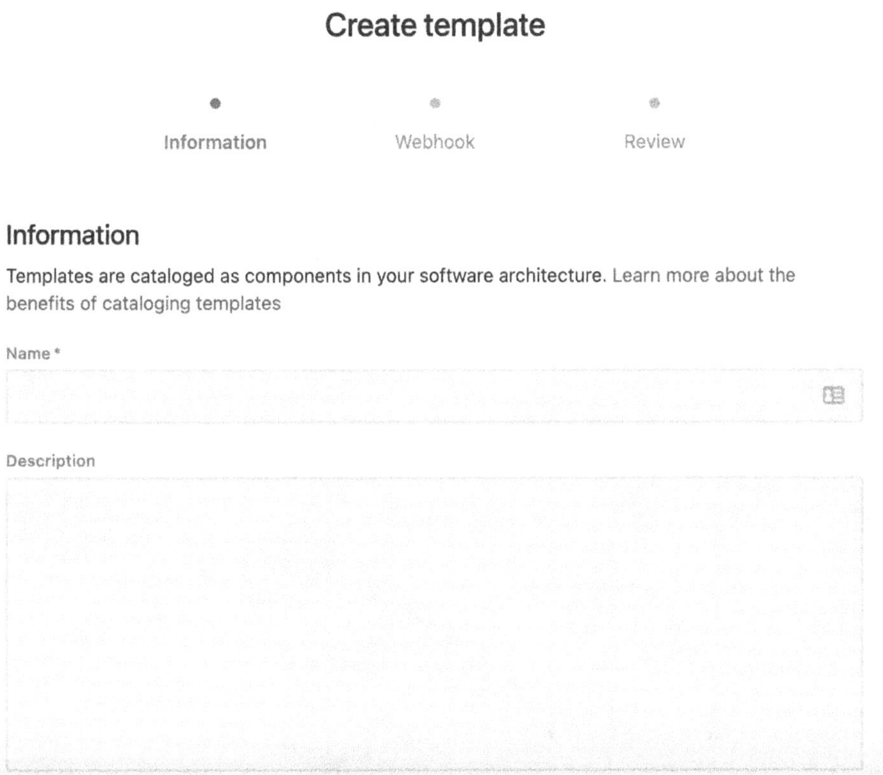

Create template

Information   Webhook   Review

Information

Templates are cataloged as components in your software architecture. Learn more about the benefits of cataloging templates

Name *

Description

Figure 11.64 – Template creation information

The only major difference is that when creating a template, you are required to provide a link to a repository. This can be any repository in your source code repository, but it needs to be connected with Compass, which is configured in the final recipe. When you create a template, your base repository is going to be forked:

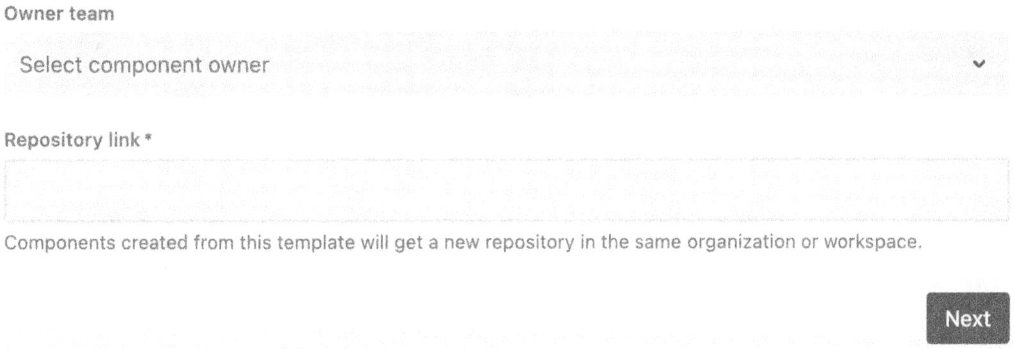

Owner team

Select component owner

Repository link *

Components created from this template will get a new repository in the same organization or workspace.

Next

Figure 11.65 – Selecting the owner and repository link

Fill out all the information and click the blue **Next** button.

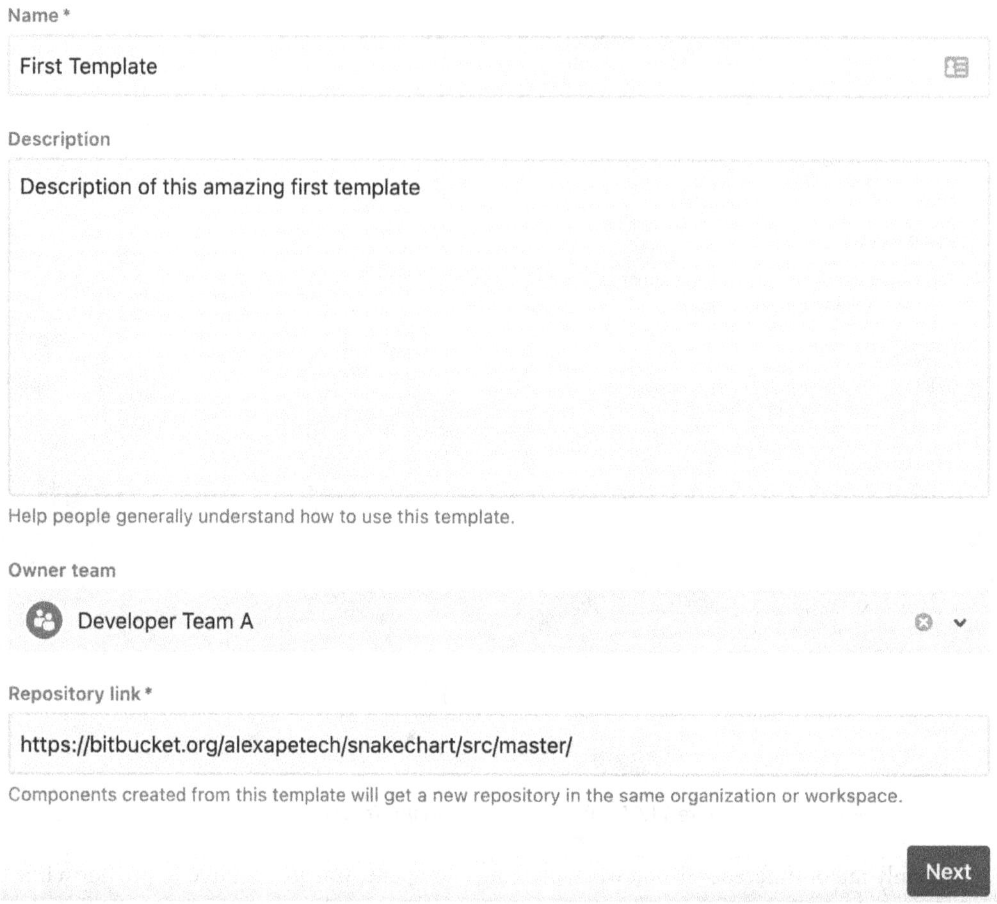

Name *

First Template

Description

Description of this amazing first template

Help people generally understand how to use this template.

Owner team

Developer Team A

Repository link *

https://bitbucket.org/alexapetech/snakechart/src/master/

Components created from this template will get a new repository in the same organization or workspace.

Next

Figure 11.66 – Completing the template creation information

3. Optionally, you can provide a webhook.

Figure 11.67 – Adding an optional webhook and parameters

4.   Review your new template and click on **Create**.

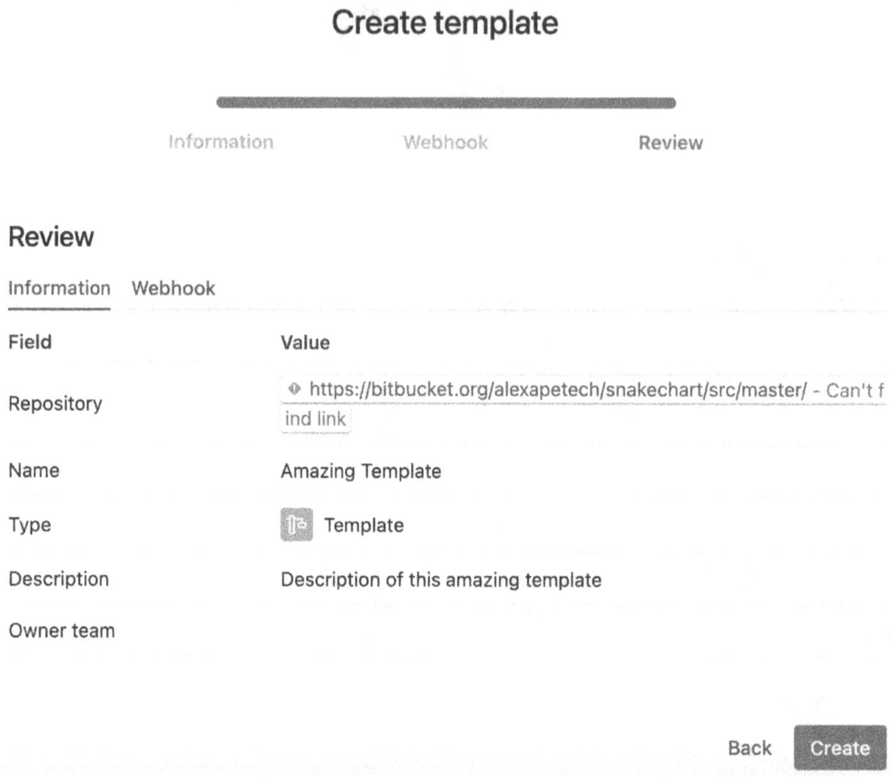

Create template

Information   Webhook   **Review**

Review

Information   Webhook

Field	Value
Repository	⊕ https://bitbucket.org/alexapetech/snakechart/src/master/ - Can't find link
Name	Amazing Template
Type	▨ Template
Description	Description of this amazing template
Owner team	

Back   Create

Figure 11.68 – Reviewing template information

Your newly created template will now be available in your component catalog.

Implementing developer CheckOps in Compass

CheckOps is an activity that teams that own the various components of their software systems perform to evaluate the health and status of their components. It is recommended to conduct a CheckOps at least once a week, but it is up to the individual teams to determine what works for them. At the very least, a CheckOps should be performed at the end of each software development sprint. Conducting a CheckOps allows teams to review scorecards and the overall health of the component(s) that are assigned to them. If a component is failing, then via the weekly CheckOps, the team can figure out a plan to address the failure and bring the component back to a healthy scorecard.

Conducting a regular CheckOps is a good idea because it allows teams to continuously monitor how the components they are responsible for are performing. Far too often, teams add components into Compass but then do not keep up with the components' health and status. Allowing for a component to fall behind or degrade in service/reliability can lead to significant software troubles later. Conducting the weekly CheckOps allows teams to consistently review key metrics and adjust on demand, as opposed to waiting until something catastrophic happens to their software. Identifying and addressing issues with a component early and often makes it so that software is of higher quality with fewer issues downstream.

Performing a weekly CheckOps is very simple and straightforward. It is all handled within the Compass application and each team can conduct their own CheckOps based on the components that are assigned to them. If components are not assigned to a team, then it is recommended that the component catalog be reviewed and a team assigned to each component as appropriate.

Up next, we'll walk you through the steps you need to follow in preparation for conducting your very first CheckOps.

How to do it...

Follow these instructions to configure Compass correctly so that your team will be able to conduct a weekly CheckOps:

1. In the Compass navigation bar, go to the **Teams** section:

Figure 11.69 – Selecting Teams from Compass navigation

2. Select your team by clicking on a team from the available teams.

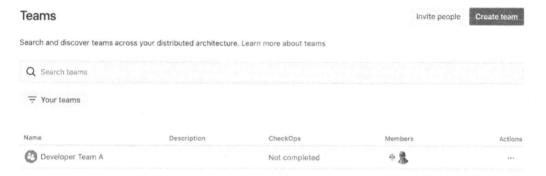

Figure 11.70 – Selecting a team from the available list of teams

3. Once you select your team, on the left navigation, select **CheckOps**.

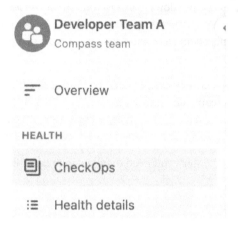

Figure 11.71 – Selecting CheckOps once a team has been selected

4. Since this will be your very first CheckOps, you'll be able to click on the blue **Start CheckOps** button.

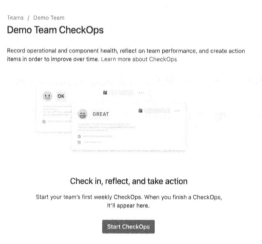

Figure 11.72 – Dashboard in Compass to initiate a CheckOps

5. This will redirect you to the **Health Details** where you'll be able to conduct your weekly CheckOps, which is covered in the next subsection.

Before you conduct your very first CheckOps, here are a few things for you to consider:

- Gain a comprehensive understanding of how your components both impact and are impacted by other elements within your software architecture.

- Record any incidents or anomalies that you may have encountered leading up to your very first CheckOps. You'll want to record these weekly as you prepare to conduct your CheckOps every week.

- Understand the service-level objectives for your component. What are the expectations that your stakeholders have for your specific component? You'll want to review this weekly to ensure that your component is meeting expectations.

- Determine a weekly cadence for your team. Every week, at the same time, your team should come together to conduct its weekly CheckOps. This should be treated similarly to any sprint planning, daily scrum, sprint reviews, and so on.

Now that you have selected your team and initiated a CheckOps, it is time to learn how to conduct the weekly CheckOps meeting.

There's more...

There is a lot more to cover, and this section will break down how you should handle your weekly CheckOps meeting.

> **Important note**
>
> As your team prepares for its weekly CheckOps, special attention should be paid to any component that has a scorecard indicating that the component needs some attention. Once you have that information and you have done some analysis/investigation into the root cause, it is time to conduct your weekly CheckOps.

Please use the following steps:

1. Within the **Health details** page, you will want to focus on the **CheckOps** section in the far-right pane within the Compass UI.

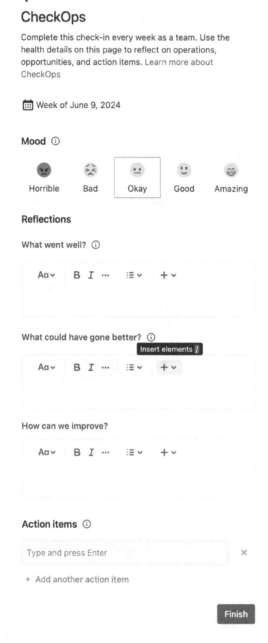

Figure 11.73 – CheckOps UI within Compass

2. Starting with **Mood**, address how your team is feeling that week? This should be with respect to the operational aspect of its components. Are the components behaving as expected, or were there multiple issues that were experienced since the last CheckOps?

    The team can choose from the available moods, which include the following:

    - **Amazing**

    - **Good**

    - **Okay**

    - **Bad**

    - **Horrible**

Figure 11.74 – Team mood selection

3. Next, there are three different reflection sections that your team will fill out together. These reflection questions are very similar to those from a traditional sprint retrospective. Let's discuss what kind of information your team should be providing for each section. You will see the following labels:

    I.   **What went well?**

         Talk about the positive things that went well for your components this past week. Sometimes, this feels like you are bragging a little bit, but it is very important that you capture the victories that your team had. It helps build a reputation and gives your team the positive reinforcement needed to keep moving in the right direction. Highlight any components that are doing extraordinarily well. Talk about all the positive things that went well and what the team/customer gained because of the items that went well.

Figure 11.75 – What went well? question

II.   **What could have gone better?**

There is always some room for improvement. In this reflection question, your team should focus on identifying items that maybe went well but could be improved. More commonly, you should capture minor problems and setbacks that the team encountered but was able to fix or resolve quickly. This is a good opportunity to find annoyances with the components and figure out ways to eliminate them. No idea is a bad idea and the team should be encouraged to surface its ideas for improvements. This often requires the team to have a high level of trust and a healthy team environment where ideas can be freely discussed.

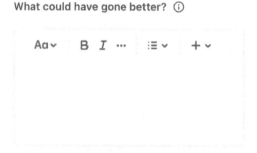

Figure 11.76 – What could have gone better? question

III.   **How can we improve?**

The last reflection question gives your team the ability to really think about how things can be better. This is similar to the previous reflection point, but this time, think about anything that needs some improvement. The improvements do not have to be technical, either. Maybe the team needs to improve its documentation or dependencies. Your team should be forward-looking, identifying opportunities where it can better support its components, while also finding ways to increase the quality of the components it is responsible for.

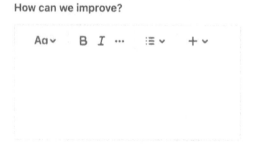

Figure 11.77 – How can we improve? question

4. The final step of the weekly CheckOps process is to capture any action items that the team needs to take within the next week or two. These action items should be reviewed every week to ensure that the team is actively working toward closing them out. Capture as many actions as the team discovers during its CheckOps. Unlike in Confluence, where you can assign the action item to an individual, the team will have to be more diligent and responsible when it comes to closing out items. You also cannot add due dates to these action items, as you can in Confluence, and the action items are not connected to Jira either.

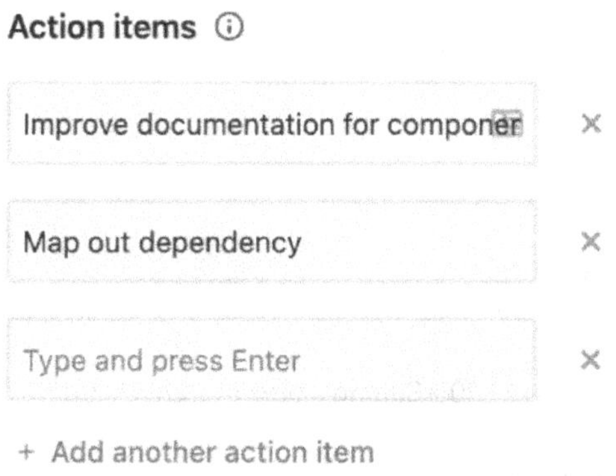

Figure 11.78 – Action items from performing CheckOps

5. Once you are finished capturing all the information, you can click on the **Finish** button.

6. After you have finished your CheckOps, you can always review it and any previous CheckOps within your team's CheckOps dashboard, which you can access by clicking on **CheckOps** in the left navigation of your team's profile in Compass.

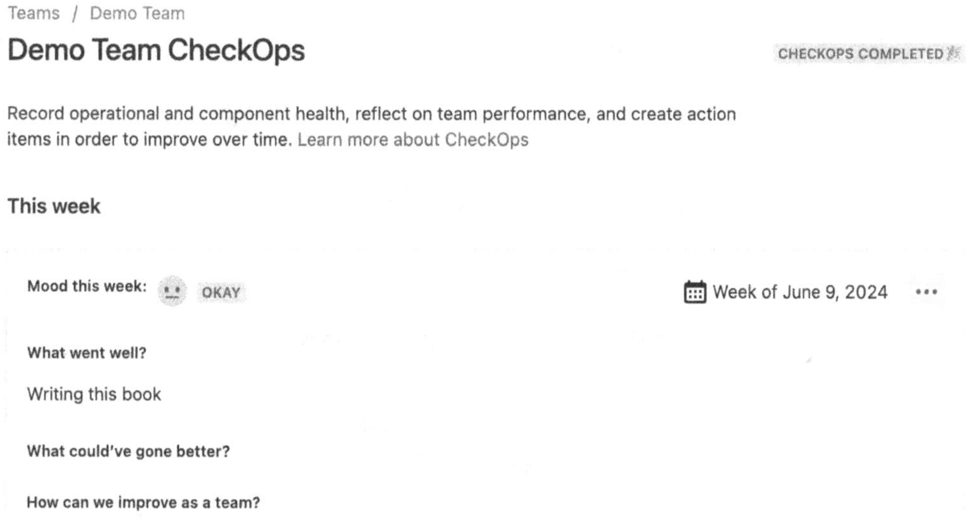

Figure 11.79 – Reviewing the previous team CheckOps

7.   From here, you edit your CheckOps by clicking on the ellipsis next to the date of the CheckOps.

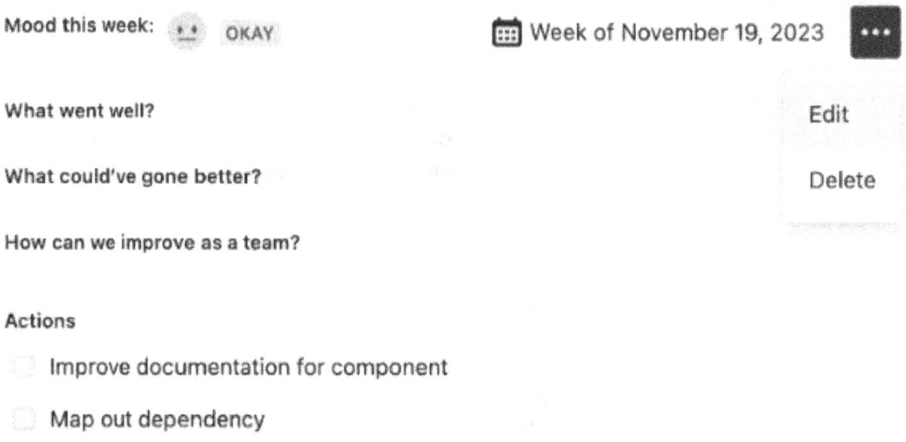

Mood this week: OKAY   📅 Week of November 19, 2023   ...

What went well?   Edit

What could've gone better?   Delete

How can we improve as a team?

Actions

☐ Improve documentation for component

☐ Map out dependency

Figure 11.80 – Editing a previous CheckOps

8.   You also have the ability to close out your actions from this view.

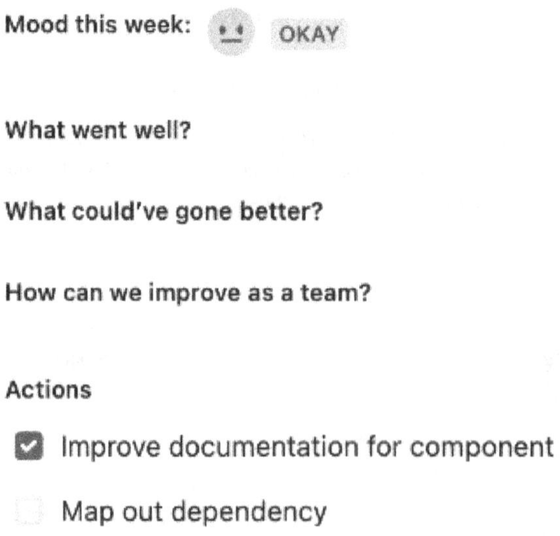

Mood this week: OKAY

What went well?

What could've gone better?

How can we improve as a team?

Actions

☑ Improve documentation for component

☐ Map out dependency

Figure 11.81 – Closing out previous actions

9. Finally, on the overview page for your team, any outstanding action items that were captured during a CheckOps session will be displayed, allowing your team to be reminded and hopefully encouraged to act on those outstanding items.

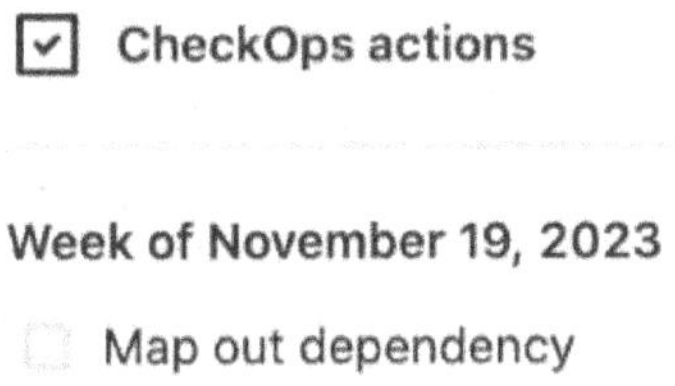

☑ **CheckOps actions**

Week of November 19, 2023

☐ Map out dependency

Figure 11.82 – CheckOps at-a-glance action review

Now that you know how to conduct your weekly CheckOps, in the next section, we'll discuss some best practices to make sure you and your team are getting the most out of your CheckOps sessions.

See also

Performing a weekly CheckOps is an easy thing to do, but there are a few recommendations that will help boost the effectiveness of your weekly CheckOps. These are the *CheckOps best practices*. By following these recommendations, you can help your team maximize the value the team and the rest of your company get from the weekly CheckOps.

Consistency

Do not skip your weekly CheckOps. Sometimes your team might not have much time, or there may be times when the scorecards are all green. Even if everything is going well, build the habit of always running your weekly CheckOps at the same time every week. If there is a holiday, make sure you schedule around the holiday. Building the habit of always running your CheckOps meeting will ensure that your team treats it as a high-priority item. Then, whenever something goes wrong, your team will already have the muscle memory of knowing how to run CheckOps sessions and will be able to focus on the critical issues as opposed to figuring out how to conduct the meeting.

Scorecard

Continuously review your scorecard and the metrics that make up your scorecard. Your scorecard is one key indicator that influences what items are discussed during the weekly CheckOps. If the scorecard is always showing positive results, but deep down, something is not working at the software level, review the items that feed into your scorecard. Make sure that they are accurate and that they appropriately tell you what is happening at the software level. Make the appropriate adjustments to your metrics that drive your scorecards to ensure that your component health reflects the most accurate data possible. This will help you have the best weekly CheckOps because you'll have the most accurate and relevant data to help drive discussions, decisions, and any actions that need to happen.

Follow through

The final recommendation is to ensure that your team always follows through on your actions. Technical debt builds up real fast if you ignore problems. You do not have to act right away, but every action that is captured during your weekly CheckOps sessions should be addressed before your team gets together again for the next CheckOps session. Failure to act is only going to result in bad metrics becoming even worse over time. An added benefit of following through and acting on the actions is that the team will observe that the actions are being resolved. This will encourage them to identify more inefficiencies and problems with the hopes that they too will be addressed.

12

Escalate Using Opsgenie Alerts

As part of Atlassian's Open DevOps package, you also get access to Opsgenie. Opsgenie is a separate product that allows teams to manage alerts, incidents, and incident escalation. It is normally used by agents using **Jira Service Management**, but it also works for software teams that need to be involved in resolving production issues. Software teams can leverage the power of Opsgenie to manage alerts and incidents that occur against the systems that have been deployed to production.

As we look at the entire DevOps lifecycle, Opsgenie facilitates monitoring activities. Once your products and services are deployed in production, monitoring and reacting to incidents is a critical piece of the feedback loop. This chapter will explore how to configure Opsgenie and integrate it with Jira.

This chapter has the following recipes:

- Setting up Opsgenie teams
- Setting up Opsgenie with Jira
- Setting up on-call schedules
- Configuring escalation policies and rules
- Escalation and notifications configuration
- Improving team communication and response with ChatOps

Technical requirements

You will need the following software:

- Jira
- Opsgenie
- Compass

Setting up Opsgenie teams

In this recipe, we are going to configure your team so that Opsgenie can send out the appropriate notifications when an incident occurs. If you have already created a team in Jira, Confluence, or Compass, you will need to recreate your team in Opsgenie because as of the time of writing, the team in Opsgenie doesn't communicate with those team settings. You will need to recreate your team in Opsgenie exactly how you have it in the other Atlassian tools if you want the same team name and members. Otherwise, you have the option to create an entirely new team in Opsgenie.

How to do it

The following steps will guide you through creating a team in Opsgenie:

1. In Opsgenie, click on **Teams** in the navigation bar.

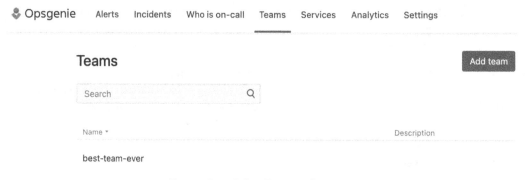

Figure 12.1 – Select Teams in Opsgenie

2. Click on **Add team**.

Figure 12.2 – Add a new team

3. Provide a name and a description of your team, and if you have the member information available, add the members' names as well. Click on **Add team** when you have finished adding your members to the new team.

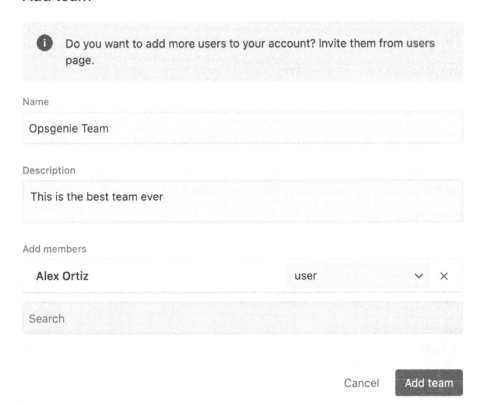

Figure 12.3 – New team details

Important note

If your team member doesn't show up, then you'll need to go to Opsgenie's settings in the navigation bar and add your users there. As noted earlier, Opsgenie is independent from the rest of the Atlassian tools and users are managed independently as well.

4. If you have not added your team members to Opsgenie, you can add them by clicking on **Members** in the left pane and then clicking **Add member** at the top of the **Members** page.

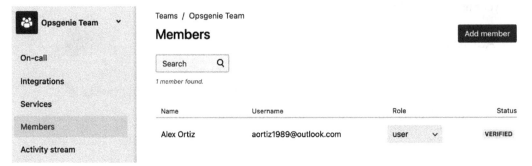

Figure 12.4 – Opsgenie settings to add new members

Add any missing team members to your Opsgenie team and assign the appropriate role (**user** or **admin**). Click **Add** when you are finished adding members.

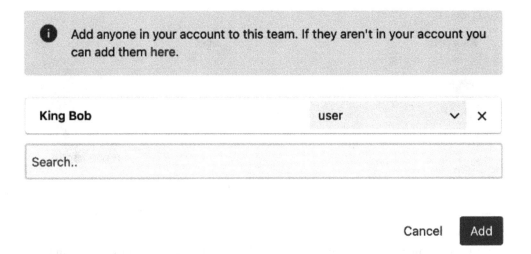

Figure 12.5 – Add a member to the team with the appropriate role

Now your team is all onboarded in Opsgenie. It is time to start creating your team's on-call schedule, which will be covered in the next recipe.

Setting up Opsgenie with Jira

Opsgenie is a separate Atlassian product that is typically bundled with Jira Service Management or when you go through Atlassian's Open DevOps package. When you sign up for either service, Opsgenie will be made available but navigating to the Opsgenie product is not very easy. For this chapter, we are going to focus on utilizing Opsgenie through Jira. Since Open DevOps does not include Jira Service Management, there will be no discussion of Jira Service Management in this chapter, but feel free to explore the capabilities of Jira Service Management as it is a great product used for IT service management teams.

How to do it

In Jira, select the Jira project your team is using to track their development tasks. Once in that project, you'll first want to enable and connect with Opsgenie. You'll need to use the following steps:

1. Select the Jira project that you want to integrate with Opsgenie, then go to **Project settings** and select **Toolchain**.

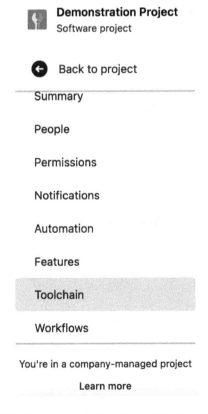

Figure 12.6 – Toolchain selection from within the Jira project

2. In the **Toolchain** window, click on **Add**. From the dropdown, select **Add on-call team**.

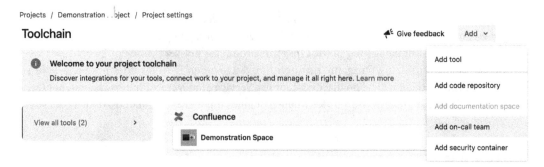

Figure 12.7 – Add new on-call team from within Toolchain

3. Select **Opsgenie** and click on **Add to project**.

Figure 12.8 – Select Opsgenie as an on-call team provider

4. Create or search for a team.

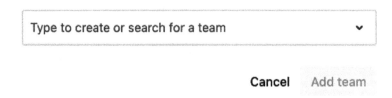

Figure 12.9 – Add an on-call team

5.   Type in the name of your team and click on the green circle to create the team.

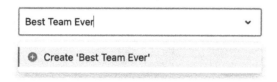

Figure 12.10 – Select the team to add

6.   Click on the **Add team** button.

Figure 12.11 – Add an on-call team

7.   Back in your Jira **Project settings**, click on **Features**.

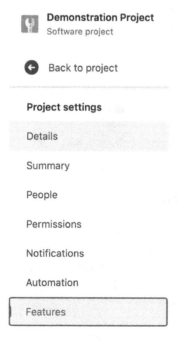

Figure 12.12 – Select Features from within the Jira project

8.   Under **Operations**, find the **On-call** section and make sure the slider is green.

Figure 12.13 – Enable On-call from within the Jira project features

You should now be able to access your on-call schedule and have a link to jump into Opsgenie from your main Jira project. In your Jira project, you'll have an **On-call** section and a link that says **View in Opsgenie**.

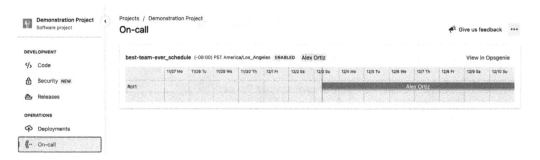

Figure 12.14 – On-call view from within the Jira project

Opsgenie is now connected with Jira and your team is now ready to start managing alerts and viewing on-call schedules from within Jira.

Setting up on-call schedules

Once your team has been created and your team members have been added, it is time to add an **on-call schedule**. The on-call schedule allows your team to divide responsibility with respect to making sure that your components and software always has coverage. If something goes wrong with a production environment, the on-call schedule will know who to notify based on the time of the incident and the team member assigned to that block of time.

How to do it

An on-call schedule in Opsgenie lets your team stay organized and distribute the responsibility for managing and ensuring that critical components are operational. The following steps will show you how to configure your team's on-call schedule:

1. In Opsgenie, click on **Teams** in the navigation bar and then click on the team for which you want to create an on-call schedule.

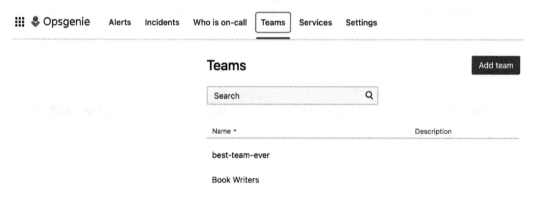

Figure 12.15 – Select Opsgenie team

2. Once you are in the team dashboard, click on the **On-call** button. This will present you with a few different options. For now, scroll all the way down and find the **On-call** section.

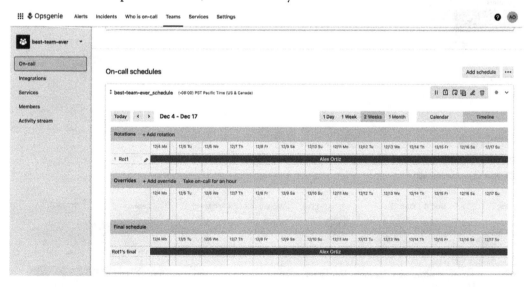

Figure 12.16 – Team On-call schedule

> **Important note**
>
> **Rotations** dictate who in your team will be on call for that specific time period. There should already be an existing rotation visible. We are going to edit this existing rotation. Optionally, you have the option to create additional rotations. Adding a new rotation can complicate things, so it is recommended that you stick to a single rotation while your team gets used to being on-call. Most software teams are not used to being on-call, so this may be a bit of a learning curve. Try to keep your rotations as simple as possible while your team is starting out. Opsgenie will be able to scale and grow with your team as you get better and more used to doing rotations. As a best practice, a teammate should be assigned at least a 1 full week as part of their rotation.

3.   Edit the existing rotation by clicking on the pencil icon button under the **Rotations** heading.

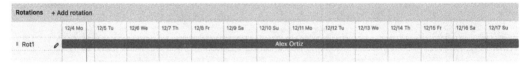

Figure 12.17 – Edit existing on-call rotation

4.   Start by changing the rotation name to anything that makes more sense for you and your team.

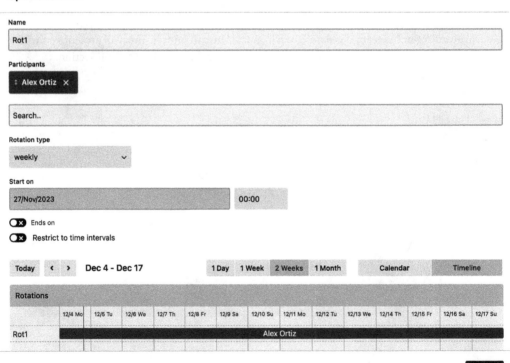

Figure 12.18 – Edit rotation name

5. From there, under **Participants**, there is a search box. There, type in the names of your team members to add them as participants of this rotation.

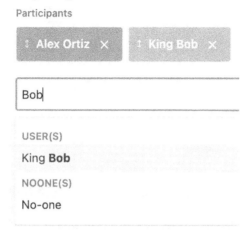

Figure 12.19 – Add team members as participants

6. Next, set your **Rotation type** option. As mentioned earlier, it is recommended you do a weekly rotation. You have the option to do **daily**, **weekly**, or a **custom** rotations.

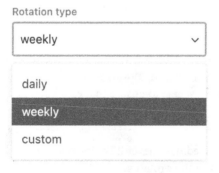

Figure 12.20 – Select rotation type

7. After that, you will set the start date for your rotation. Optionally, you can set when the rotation is going to end. Most teams will have a rotation that lives on forever.

Figure 12.21 – Select rotation start date

8.   Once you fill all of that information out, Opsgenie will create your schedule:

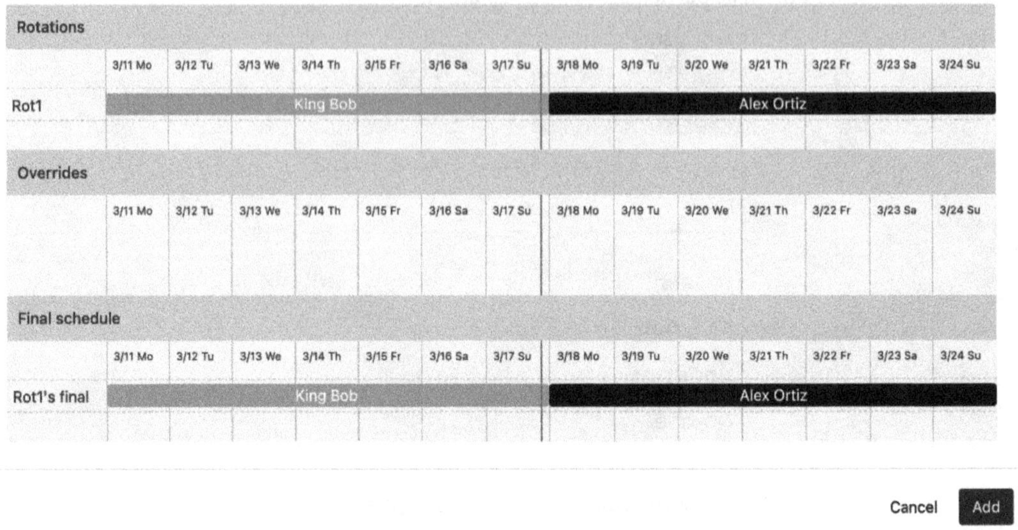

Rotations														
	3/11 Mo	3/12 Tu	3/13 We	3/14 Th	3/15 Fr	3/16 Sa	3/17 Su	3/18 Mo	3/19 Tu	3/20 We	3/21 Th	3/22 Fr	3/23 Sa	3/24 Su
Rot1				King Bob							Alex Ortiz			

Overrides														
	3/11 Mo	3/12 Tu	3/13 We	3/14 Th	3/15 Fr	3/16 Sa	3/17 Su	3/18 Mo	3/19 Tu	3/20 We	3/21 Th	3/22 Fr	3/23 Sa	3/24 Su

Final schedule														
	3/11 Mo	3/12 Tu	3/13 We	3/14 Th	3/15 Fr	3/16 Sa	3/17 Su	3/18 Mo	3/19 Tu	3/20 We	3/21 Th	3/22 Fr	3/23 Sa	3/24 Su
Rot1's final				King Bob							Alex Ortiz			

Cancel   Add

Figure 12.22 – Updated rotation on-call schedule

Click on the **Add** button and your on-call schedule will be set.

> **Important note**
>
> At this point, you can make additional on-call schedules for weekends, holidays, and pre-planned days off. Whatever your team may need, Opsgenie is flexible enough and you will be able to put your team's schedule in Opsgenie to ensure that escalation policies and routing rules in the next recipe work effectively.

Your on-call schedule is now created. Whenever it is someone's turn to be on call, they will receive a notification from Opsgenie to start their rotation.

Configuring escalation policies and rules

Once your on-call schedule has been configured, it is time to configure the escalation policies and routing rules. Opsgenie uses the **escalation policies**, **routing rules**, and the **on-call schedule** from the previous recipe to determine who will be alerted when an incident occurs.

> **Important note**
>
> It's important to get the escalation policies and routing rules correct otherwise you risk sending the alert to the wrong team. The escalation policy you configure will determine which users are notified when an alert occurs. Most importantly, the escalation policy determines the order of which users get notified and when based on how much time has elapsed since the alert was created. If your team doesn't reply, the next person in the chain gets notified until someone does something to resolve the alert/incident. The routing policy will determine how the notification gets routed when an alert occurs. There are a few different options available to you which will be explored in more detail in this recipe.

How to do it

By default, you should already have an escalation policy created in Opsgenie. Similar to creating the on-call schedule, you have the option to modify the existing escalation policy, or you can add additional escalation policies.

The escalation policy determines who gets notified of any alerts in Opsgenie when something goes wrong. Let's configure your first escalation policy:

1. Select your team from Opsgenie.

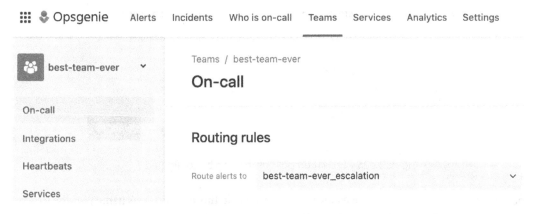

Figure 12.23 – Selected team to configure escalation policy

2. Start by hovering over the existing escalation policy and you will see the **Edit** button reveal itself.

Figure 12.24 – Edit existing escalation policy

Clicking on the **Edit** button will pull up a menu that allows you to configure your escalation.

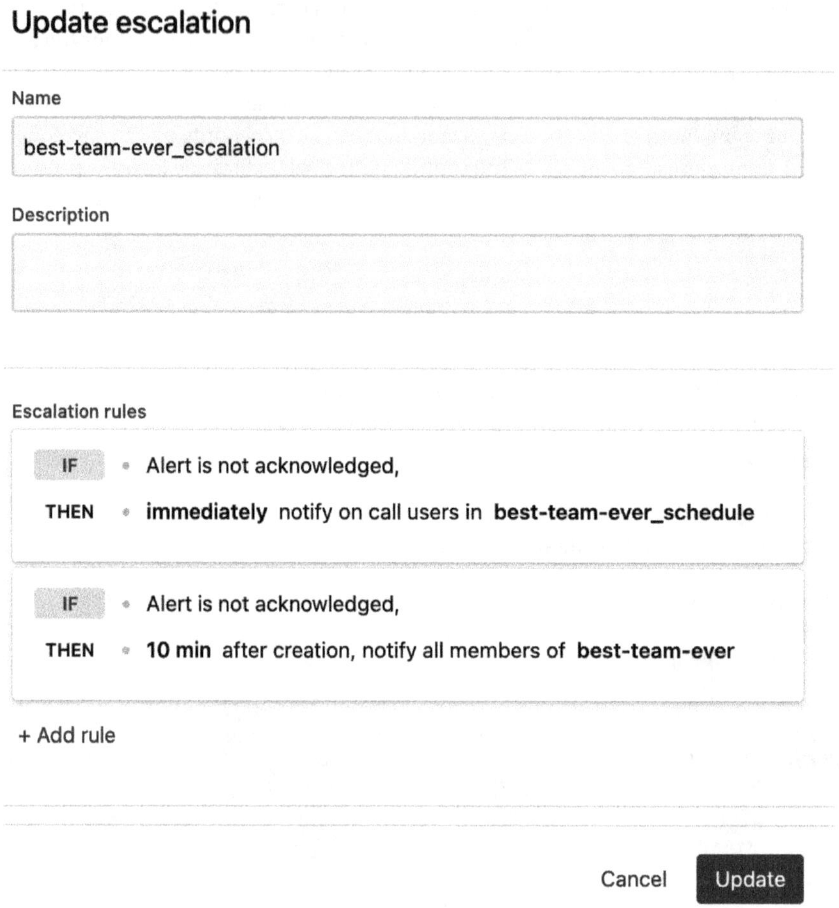

Update escalation
Name
best-team-ever_escalation
Description
Escalation rules
IF
Alert is not acknowledged,
THEN
immediately notify on call users in best-team-ever_schedule
IF
Alert is not acknowledged,
THEN
10 min after creation, notify all members of best-team-ever
+ Add rule
Cancel
Update

Figure 12.25 – Escalation policies that can be configured

3. The first item to change is the **Name** setting. By default, this is your team's name plus the word `escalation`. Feel free to change it to whatever you want.

Name
Updated best-team-ever_escalation

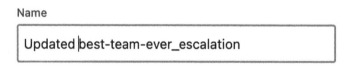

Figure 12.26 – Updated escalation name

4. Next, you can optionally provide a **Description** text to help your team have a better understanding of what the purpose of this specific escalation policy is.

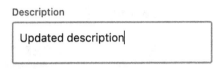

Figure 12.27 – Updated escalation description

5. Finally, you can modify the existing rules or add additional rules.

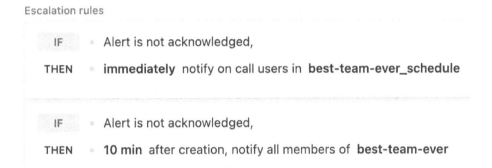

Figure 12.28 – Updated escalation rules

> **Important note**
> By default, the escalation policy will immediately notify all the members of your on-call schedule as soon as an alert is created, and it is not acknowledged right away. Then, after 10 minutes, if no one from the on-call schedule has acknowledged the alert, a notification is sent to all the members of your team. This will include all the members not currently on the on-call rotation.

6. If you need to escalate things even further, you can add additional rules.

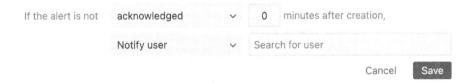

Figure 12.29 – Additional escalation rules

7. When adding an additional rule, you first need to determine whether the escalation policy will escalate when an alert is not acknowledged or not closed.

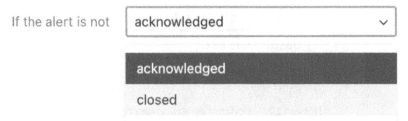

Figure 12.30 – Determine when rule is triggered

8. Once you make that decision, then you pick how much time can elapse after the alert is created.

Figure 12.31 – Determine how long after trigger alert should wait

9. Next, pick who will receive the notification of the escalation. You have quite a few options and just keep in mind that as time elapses, you want to increase the visibility of the alert to the appropriate people that can then take action to address the alert.

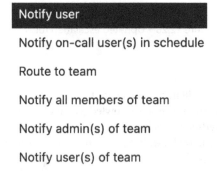

Figure 12.32 – Select who should get notified of the alert

10. Finally, pick a specific team or user to receive the notification. The teams available will be determined by the previous recipe.

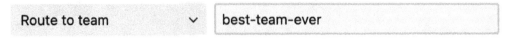

Figure 12.33 – Select the team that will receive the alert

11. When you are finished making the escalation rule, click on **Save**.

Figure 12.34 – Save escalation rule

There's more

Now that the escalation policy exists, it is time to work on the routing rules.

Routing rules

Like with the escalation policy, each team should have a default routing rule in place. Routing rules are quite simple, and you can either modify the existing routing rule or add a new routing rule altogether. Let us use the following steps:

1. Back in Opsgenie, under the **On-call** section, we are going to modify the default routing rule.

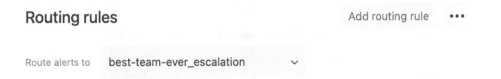

Figure 12.35 – Routing rule configuration

> **Important note**
> You have three main options when it comes to setting up your routing rules. You can route the alert to no one, you can route it to your escalation policy, or you can route it directly to your on-call team.

2. Click on the existing routing rule and then you will be able to select which type of route you want to route your alerts to. The following screenshot shows the options that are available.

Figure 12.36 – Select escalation routing rule

3. Alternatively, you can create a new routing rule if you do not want to use the default one that Opsgenie provides. Creating a routing rule presents you with a few more options when compared to creating an escalation rule. The following steps will explain how to configure a routing rule:

I.   To get started, click on **Add routing rule** in the top-right corner.

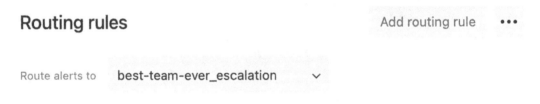

Figure 12.37 – Add a new routing rule

II.   After you click on the **Add routing rule** button, you'll be prompted to provide a name for your routing rule.

Add routing rule

Name

Figure 12.38 – Provide the routing rule a name

III.   Next, you'll pick when the routing rule goes into effect. You can pick from **Match all alerts**, which means that for any alert, the routing rule will go into effect (default). Alternatively, you can make your routing rule trigger based on the condition of the alert.

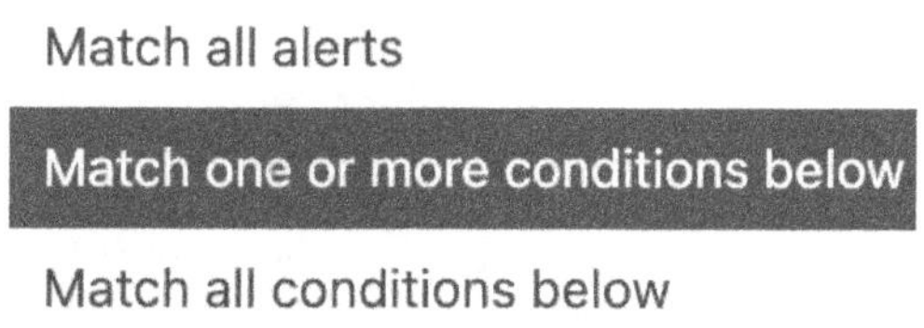

Match all alerts

Match one or more conditions below

Match all conditions below

Figure 12.39 – Select what conditions trigger the routing rule

IV.   You can create conditions based on either the alert's priority or a tag associated with it. The most important takeaway here is that you can make the routing rule go into effect if an alert with a specific priority is received.

If the following conditions are met

Match one or more conditions below ˅

Priority ˅   --   ˅   Equals ˅   P3-Moderate ˅

+ Add new condition

Figure 12.40 – Determine conditions for the routing rule

V.   Once you set up your conditions, then the final step is to determine the action that is going be triggered by the automation rule.

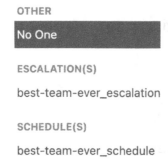

Figure 12.41 – Determine the action the routing rule will take

VI.   Once you pick the route, simply click on the **Add** button to finish creating your new routing rule.

Figure 12.42 – Add the new routing rule

Now that your escalation and routing rules are configured, it is time to manage the notifications that are sent out from Opsgenie.

Escalation and notifications configuration

When an incident occurs in a production environment, an alert can be raised in Opsgenie. You do this by opening Opsgenie and creating an alert directly from within Opsgenie. When Opsgenie is bundled with Jira Service Management, the creation of an alert can happen from within a support request. However, since Jira Service Management is not part of the Open DevOps package, an alert must be manually created in Opsgenie.

How to do it

Let's use the following steps to create an alert:

1.   To create an alert, click on **Alerts** in the navigation bar within Opsgenie.

Figure 12.43 – Select Alerts from the Opsgenie menu

2. Once you are in the **Alerts** section, click on the **Create alert** button.

Figure 12.44 – Create an alert in Opsgenie

3. Fill out the information for the alert.

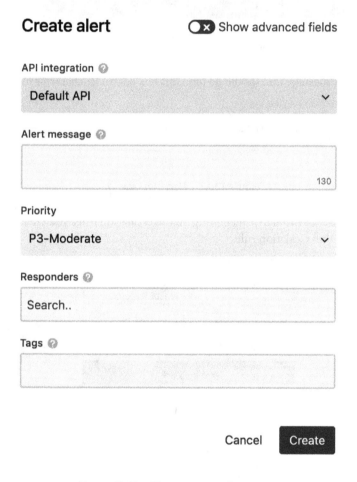

Figure 12.45 – Alert creation information

4. Select the API integration you would like to utilize for this alert. **Default API** is recommended for beginners and is the only option if you are in the free version of Opsgenie.

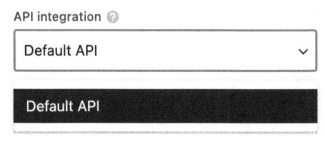

Figure 12.46 – Alert API selection

5. Enter the **Alert message** that should go out. This box has a 130-character limit.

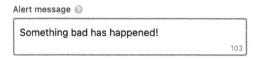

Figure 12.47 – Alert message

6. Select the priority of the alert. Depending on the escalation rules you created, this alert will impact/trigger that escalation rule.

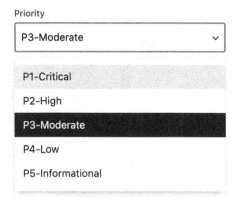

Figure 12.48 – Alert priority

7. If you know who the responders of the alert should be, fill that out. Otherwise, the routing rules will figure out who should be notified based on the values you configured in your routing rule configuration.

Figure 12.49 – Alert responders

8. You can leverage **Tags** if your routing rules are configured to take advantage of tags instead of priority.

Figure 12.50 – Alert responders

9. Finally, click on **Create** to create the alert.

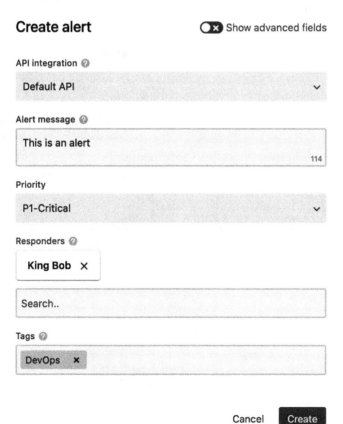

Figure 12.51 – Complete alert example

The alert will now be visible in the **Alerts** section.

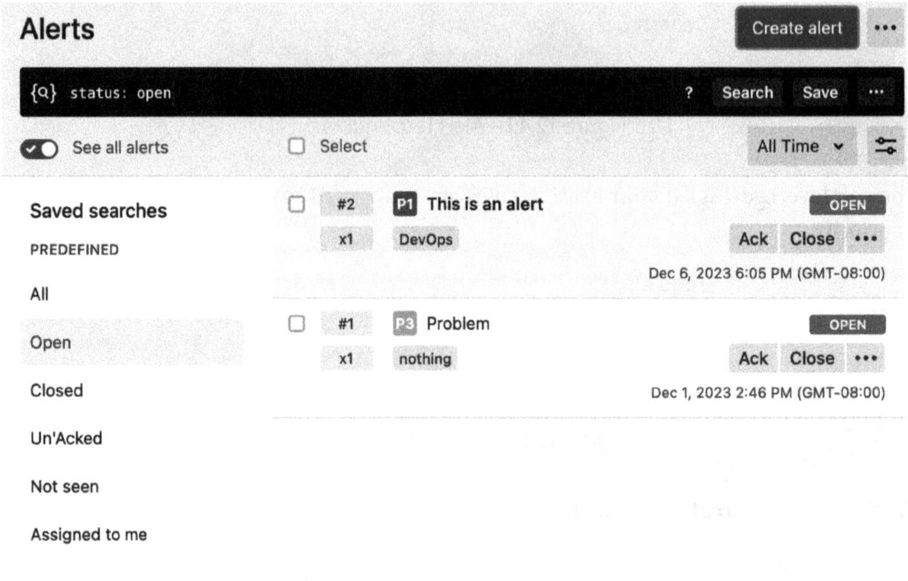

Alerts

Create alert ...

{Q} status: open   ? Search Save ...

See all alerts   Select   All Time ⌄

Saved searches   #2 **P1** **This is an alert**   OPEN
                          x1   DevOps   Ack Close ...
PREDEFINED
                                                            Dec 6, 2023 6:05 PM (GMT-08:00)
All
                          #1 **P3** Problem   OPEN
Open   x1   nothing   Ack Close ...

Closed   Dec 1, 2023 2:46 PM (GMT-08:00)

Un'Acked

Not seen

Assigned to me

Figure 12.52 – View of available alerts

Important note

Based on your routing rules and escalation policy, each alert will trigger the appropriate notifications so that your team can triage and fix the problem that triggered the alert.

Once the alert has been acknowledged, the notifications will stop going out. This means that your team has addressed the problem or has determined that an incident has occurred and the alert needs to be escalated to an incident.

If you click on the alert, you'll be able to see all the automatic notifications that were sent out for the alert.

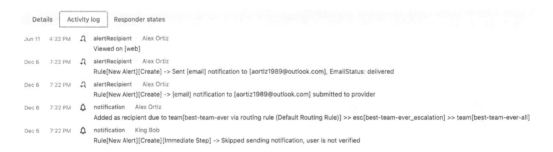

Figure 12.53 – Alert history

Alerts are critical to ensure your team is notified of critical events and know when they need to engage to fix a problem. Make sure you refine your alerts as your ecosystem changes to ensure your team is appropriately notified of problems when they first occur.

Improving team communication and response with ChatOps

When issues arise, your team needs to communicate quickly. In the digital age, relying on emails is not going to cut it when seconds matter. The best way for teams to engage is with instant communication. Most teams today already utilize popular communication tools such as Slack and Microsoft Teams. Opsgenie integrates with both of those tools, which then allow your teams to quickly communicate when an alert or incident occurs.

While Opsgenie integrates with a variety of instant messaging tools, this recipe is going to focus on getting Opsgenie integrated with Slack. Integrating with Slack not only allows your team to communicate quicker, but also enables your team to get alert information, on-call schedules, and so much more from within the Slack interface.

How to do it

Integrating Opsgenie with Slack is straightforward. The following steps will walk you through everything you need to know to get this integration configured:

1. To initiate the integration with Slack, start by going to the **Teams** section within Opsgenie and then select your team.

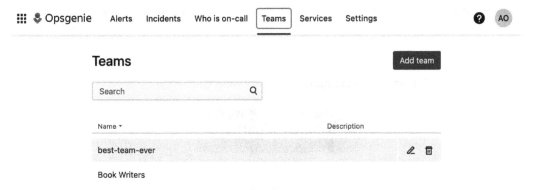

Figure 12.54 – Available teams in Opsgenie

2. Once you are in your team dashboard, click on **Integrations** on the left-hand side.

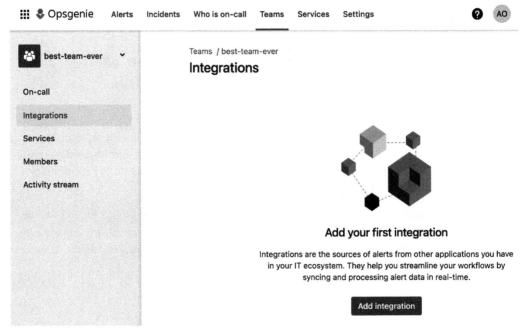

Figure 12.55 – Select integrations for your selected team

> **Important note**
> Please note that if you are on the free version of Opsgenie, this is the only way to integrate with Slack. There is a different method through Opsgenie's settings page, but that is only possible for teams that are paying for Opsgenie.

3. Click on **Add integration** to get started.

Figure 12.56 – Select Add integration to get started

4. Select **Slack**.

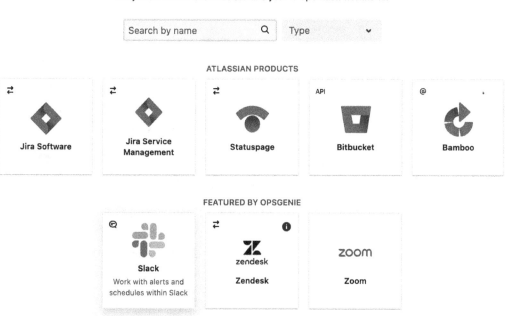

Figure 12.57 – Available Opsgenie integrations

5.   Follow the instructions displayed. You will need to have a channel ready on Slack before you continue.

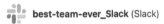 **best-team-ever_Slack** (Slack)

Real-time messaging, archiving and search.

Opsgenie has bi-directional integration with Slack. Opsgenie alert activity (alert created, acknowledged, closed, etc.) is posted to rooms in Slack. Users can use Slack buttons and /genie command to interact (ack, add not, assign, close, etc) with Opsgenie alert from Slack chat rooms.

Instructions to setup Opsgenie Slack App   ⌄

best-team-ever   ⌄   🟦 Add to Slack

- Click "Add to Slack" button.

- Select "Slack Team" and "Channel".

- Click "Authorize" button to give access to Opsgenie Slack App.

- Integration will be automatically created. (If you are not logged into Opsgenie, you will need to log in.)

- Note that you need to have required permission to add integration in Opsgenie.

- For more information, please refer to the support document..

Figure 12.58 – Slack integration with Opsgenie instructions

6. After you pick your channel in Slack, confirm the settings that you want enabled in Opsgenie. Once you have selected the options that you are satisfied with, click on **Save integration**.

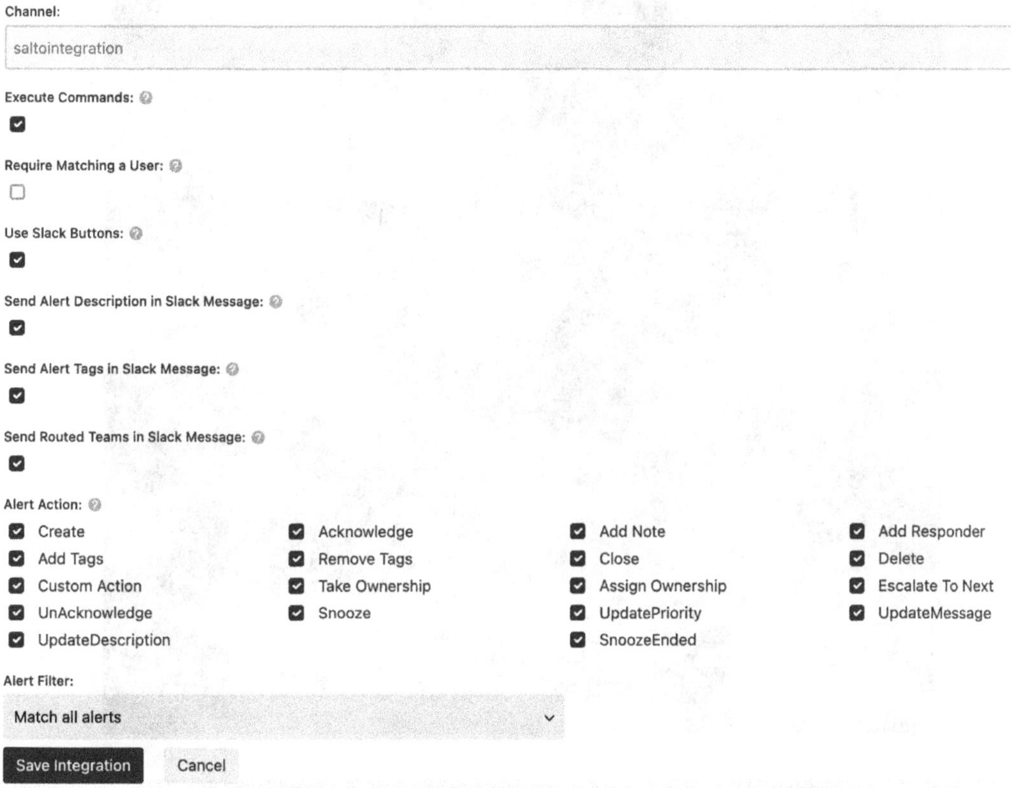

Channel:

saltointegration

Execute Commands: ⓦ
☑

Require Matching a User: ⓦ
☐

Use Slack Buttons: ⓦ
☑

Send Alert Description in Slack Message: ⓦ
☑

Send Alert Tags in Slack Message: ⓦ
☑

Send Routed Teams in Slack Message: ⓦ
☑

Alert Action: ⓦ

☑ Create	☑ Acknowledge	☑ Add Note	☑ Add Responder
☑ Add Tags	☑ Remove Tags	☑ Close	☑ Delete
☑ Custom Action	☑ Take Ownership	☑ Assign Ownership	☑ Escalate To Next
☑ UnAcknowledge	☑ Snooze	☑ UpdatePriority	☑ UpdateMessage
☑ UpdateDescription		☑ SnoozeEnded	

Alert Filter:

Match all alerts   ⌄

Save Integration   Cancel

Figure 12.59 – Select configurations for Slack integration

7. Once the integration is configured, you can test the configuration by going to Slack and then typing in the /genie command.

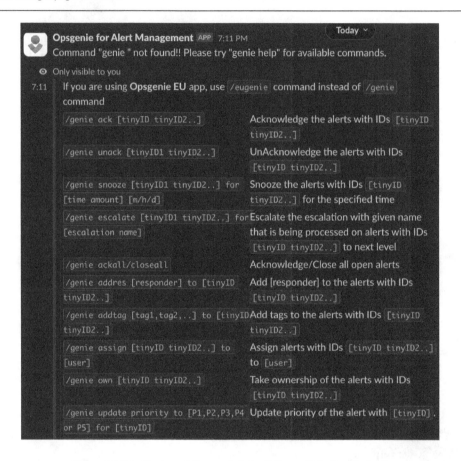

Figure 12.60 – Slack and Opsgenie integration within the Slack UI

8.   Finally, create a sample alert and you'll see the alert show up in Slack. From within Slack, you can now address your alerts without having to disrupt your workflow or team communication.

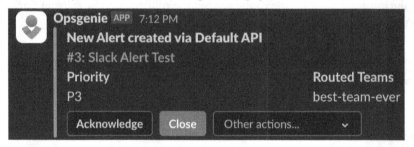

Figure 12.61 – Opsgenie alert within Slack

Slack is now integrated with Opsgenie and your team will now be able to receive Opsgenie notifications within Slack, which is very helpful since your team will not have to change tools.

Part 4:
Putting It into Practice

We now take everything we've learned in the book so far and apply it. We go through a sample project where you start with an idea in Jira Product Discovery, create an issue in Jira, and create a branch for development in Bitbucket.

We then perform a commit to look at the pipeline execution from testing to deployment. During this execution, we can see the results in both Bitbucket and Compass.

We conclude our journey with a look into the future, where artificial intelligence is already changing things. We share some final thoughts and wish you well.

This part has the following chapters:

- *Chapter 13, Putting It All Together with a Real-World Example*
- *Chapter 14, Appendix Key Takeaways and the Future of Atlassian DevOps Tools*

13

Putting It All Together with a Real-World Example

Let's take what we've learned in the previous chapters and apply this from start to finish.

Let's say you are an employee at **DevOps Products Inc. (DPI)**. You are responsible for using the toolchain by using Atlassian products and third-party products to optimize your development for the flagship product, DPI-Sync, which connects to Discord servers and sends messages. How would you begin?

To accomplish this, we are going to simulate this process by completing the following recipes:

- Creating an idea in **Jira Product Discovery (JPD)**
- Connecting an idea to an epic in Jira
- Creating a story in Jira
- Creating a code change in Bitbucket
- Committing changes in Bitbucket/Start Bitbucket Pipeline Build
- Executing Snyk scanning through Bitbucket Pipelines
- Displaying the Bitbucket Pipelines build status in Compass
- Deploying Bitbucket Pipelines
- Creating an Opsgenie alert for the Jira project
- Creating a bugfix branch
- Committing a bugfix and watching the pipeline's execution

Remember that more detailed explanations of these recipes can be found in the preceding chapters of this book. For now, let's learn how to get this sample project going.

Technical requirements

For this chapter, you'll need to install and configure the following Atlassian products:

- JPD
- Jira
- Bitbucket
- Compass
- Opsgenie

The sample code for this chapter can be found in the `Chapter13` folder of this book's GitHub repository (`https://github.com/PacktPublishing/Atlassian-DevOps-Toolchain-Cookbook/tree/main/Chapter13`).

Creating an idea in JPD

We'll start the development process by deciding what to build. We can gather various inputs from our competitors, our customers, and other sources to generate ideas for new features and products to develop.

We will store our **idea** for a new feature in JPD.

Getting ready

In our **JPD** project, we must work with the Jira administrator and the Jira project administrator to ensure that we have the appropriate role to create a JPD idea.

Your project administrator can ensure access to create ideas by performing the following steps:

1.   Select the **Project settings** option from the sidebar.

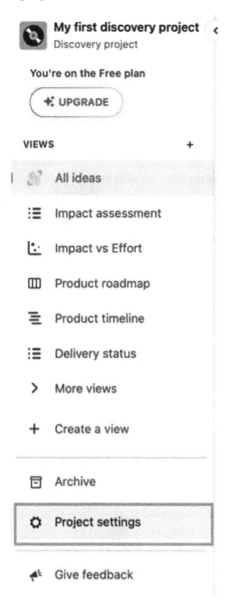

Figure 13.1 – Selecting Project settings

2. From the **Project settings** sidebar options, expand **Features** and select **Create ideas**. The page that appears should look like this:

Projects / My first discovery project / Project settings
Create ideas

Allow contributors to create ideas
Contributors can't create ideas in this project.   Preview in project ◉   Learn more ›

Figure 13.2 – Create ideas

3. Enable **Allow contributors to create ideas** by switching the toggle on.

Note that while enabled, contributors can create ideas but cannot edit them or delete them. For any idea, contributors can add comments, attachments, and insights.

How to do it...

We've already covered how to create ideas either by using the **Create** button or from within a view. Let's review how to create a new idea by using the **Create** button:

1. Click the blue **Create** button on the navigation bar.

2. A modal will appear that shows the standard fields needed for the idea. Fill in those fields and click the **Create** button.

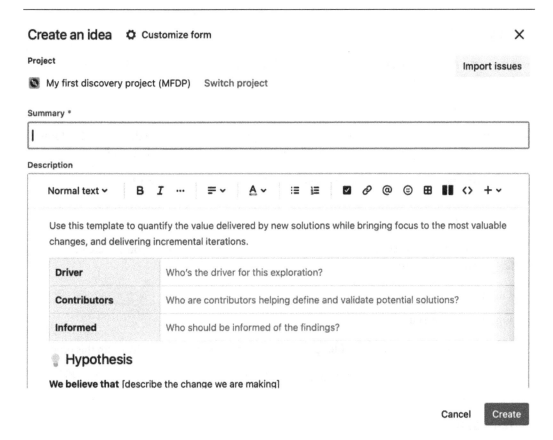

Figure 13.3 – Create an idea

3. The Jira admin can change what fields appear on the modal by creating forms. This may also include pre-filling the **Description** field by using a template.

We now have an idea that we can use to track desirability and customer sentiment by adding insights and other information. The next step is to create a Jira issue to track development work and link it to the original idea.

Connecting an idea to an epic in Jira

At this point, we're starting to move from ideation to implementation. This will require us to move from JPD to Jira. Let's see how the handoff is done.

Getting ready

As we saw in the *Delivering ideas for development in Jira* recipe in *Chapter 2*, before creating an **epic**, please verify the following:

- There is at least one Jira project available to contain the epic.

- The person creating the epic has the appropriate permissions to create the epic in the target project. This same person should be a member of the JPD project or a Jira admin.

When both conditions are met, it's easy to create an epic as a delivery ticket in JPD.

How to do it...

In the *Delivering ideas for development in Jira* recipe in *Chapter 2*, you saw that you can create an epic from the following two locations:

- On the idea's page

- On the **Idea** panel in a view

Let's learn how to create an epic from the **Idea** page so that we can familiarize ourselves with this process:

1. On the **Idea** page, click on the **Delivery** tab.

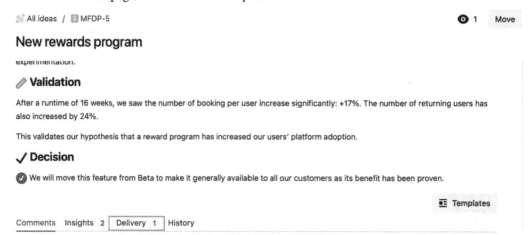

Figure 13.4 – Selecting the Delivery tab on the Idea page

2. The **Delivery** tab will expand, giving you two options:

   - Create a new delivery ticket (epic)

   - Create a link to an existing delivery ticket

Selecting **Create a delivery ticket** displays text areas for the project, including areas for specifying the **issue type** and **summary**. These fields are typically prefilled, as shown here:

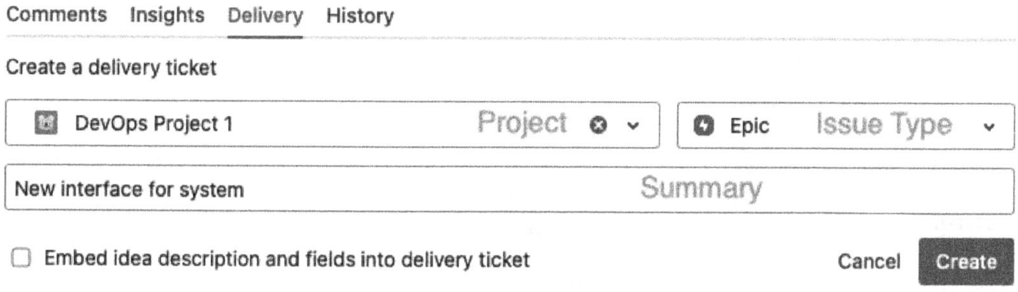

Comments   Insights   Delivery   History

Create a delivery ticket

| 🎮 DevOps Project 1 | Project ⊗ ∨ | | ⊕ Epic | Issue Type ∨ |
| New interface for system | | Summary | | |

☐ Embed idea description and fields into delivery ticket   Cancel   **Create**

Figure 13.5 – Create a delivery ticket

3.   Select **Project** and **Issue type** values and fill in the desired **Summary** before clicking **Create**.

We now have an epic in Jira that captures the feature described by our idea in JPD.

Our development team cannot work on the entire scope of an epic in one go. Refining our epic by creating *stories* that can describe small bits of functionality that can be worked on to gradually accumulate the total functionality of our feature is the best way to move forward. So, let's learn how to do this.

Creating a story in Jira

In Jira, we can use epics as containers for the small individual pieces of functionality that are captured as stories. Our epic will refer to these smaller stories as its children.

Let's create child stories directly from the epic.

Getting ready

Similar to the previous recipe, we require Jira to be installed and configured so that the following conditions can be met:

- There is at least one Jira project available to contain the story and it is connected to the epic.
- The person creating the story has the appropriate permissions to create the epic and story in the target project. This same person should be a member of the Jira project or a Jira admin.

When both conditions are met, it becomes easy to create the epic and any child stories.

How to do it...

You can create a child story that links to the epic at two locations:

- On the page for the epic
- On the **Timeline** view

Let's look at creating child stories from each location.

Creating a child story from the epic page

Epics are generally too large for a single development team to handle in a single Sprint. As the saying goes, you eat an elephant one bite at a time.

In Jira, this is done by creating child stories. Let's see how to do this from the epic page.

1. One of the ways to get to an epic page is to select the **Issues** page from the **Project** sidebar.

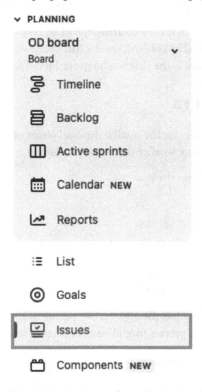

Figure 13.6 – Selecting Issues from the Project sidebar

2. Set the filters of interest to limit the search. When you see the epic, click on its **Key** or **Summary**.

3. Once on the epic page, click on the **Add a child issue** button.

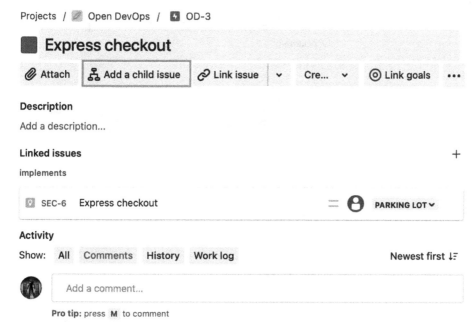

Figure 13.7 – Add a child issue

4. In the **Child issues** section, select the issue type (**Story** for user stories), fill in a summary in the **What needs to be done** area, and click **Create**.

Figure 13.8 – Creating a child issue

The parent epic now has a story linked as a child.

Creating a child story in the Timeline view

The **Timeline** view shows the issues of a project laid out against time to determine the overall progress of all the work that a project does. An epic may be placed in the timeline with an expected start and due date. Child stories can be created and planned for periods between an epic's expected start and due date.

To create a child story of an epic, perform the following steps in the **Timeline** view of your Jira project:

1. To get to the **Timeline** view, select **Timeline** in the **Project** sidebar.

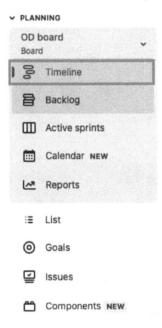

Figure 13.9 – Selecting Timeline

2. Hover over the epic until a plus sign (+) appears.

3. Select the issue type from the second dropdown and fill in the **What needs to be done?** area to populate the Summary. Press *Enter* when you're finished.

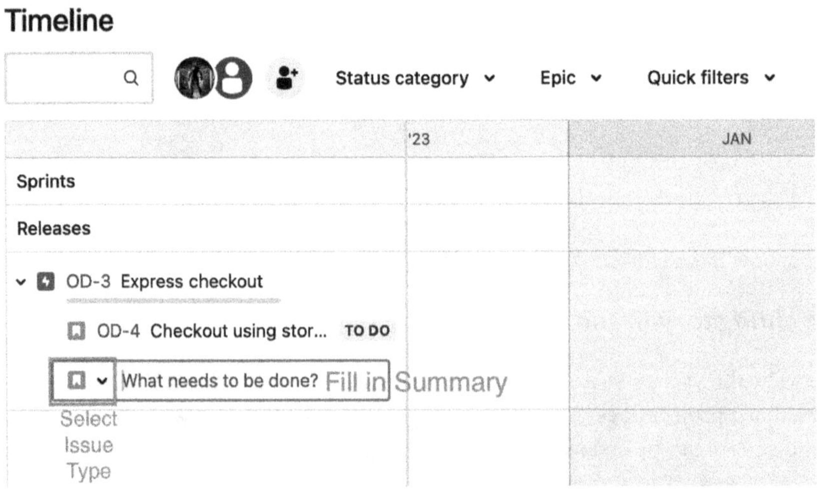

Figure 13.10 – Entering a child story in the Timeline view

A child story will now appear below the parent epic.

Creating a code change in Bitbucket

At this point, we are ready to begin developing our story in Jira. Let's say you're on a Scrum team and the story was selected for the upcoming sprint or that capacity opened up on your Kanban team to move the story from the backlog. Regardless, it's time to start development.

This recipe will have us move from Jira, where the story is planned and designed, to a Git-based tool, where we will create and store our implementation. This is where our **continuous integration and continuous Deployment (CI/CD)** pipeline will start.

While the instructions here apply to Jira connected to a Bitbucket repository, this will work for any Git-based server tool, such as GitHub or GitLab, as we saw in *Chapter 4*.

Getting ready

One thing we need to make sure we have for this recipe is a Bitbucket workspace that's been created with a Bitbucket repository. This process was explained in *Chapter 5*. There, we set up Bitbucket by performing the following steps:

1. First, we created a Bitbucket workspace. If also desired, we created a project.

2. Then, we created a Bitbucket repository.

Now that we've created our workspace and repository, we need to connect it to our Jira project. We learned how to connect our Jira project to a Bitbucket repository in *Chapter 1*. There, we performed the following steps:

1. First, we selected **Tool Chain** under **Project settings** as a Jira project administrator.

2. Then, we selected the **Build** section on the **Toolchain** page.

3. On the **Toolchain** page, we clicked **Add**.

4. From here, we selected **Bitbucket** as the repository provider.

5. Finally, we selected the repository to add and clicked **Add repository**.

We are now ready to start development with a Jira project connected to a Bitbucket repository.

How to do it...

To begin development, we must create a branch in Git to place new or changed files. If our Git repository is connected to our Jira project, we can create the branch directly from Jira and the branch will be linked to the issue describing the context.

To create this branch from the Jira story, perform the following steps:

1. Look for the **Development** section of the Jira issue. This set of fields notes the branches, commits, and pull requests associated with the Jira issue. The following screenshot shows where we can find these **Development** fields:

Figure 13.11 – Development information

2. To create a branch, click the **Create branch** link. You'll be taken to a page that allows you to configure the branch of the repository, the type of branch, and the branch's name. This can be seen in the following screenshot:

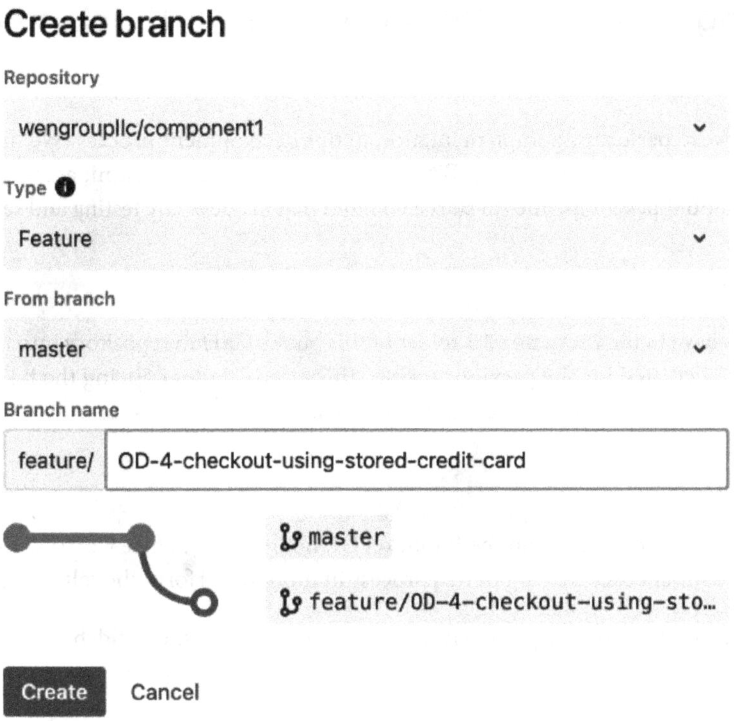

Create branch

Repository

wengroupllc/component1   ⌄

Type ⓘ

Feature   ⌄

From branch

master   ⌄

Branch name

| feature/ | OD-4-checkout-using-stored-credit-card |

ᛘ master

ᛘ feature/OD-4-checkout-using-sto…

Create   Cancel

Figure 13.12 – Create branch

Click **Create** to create your branch. Your branch will be created on Bitbucket Cloud with instructions on performing `git checkout` on your local repository copy, as shown here:

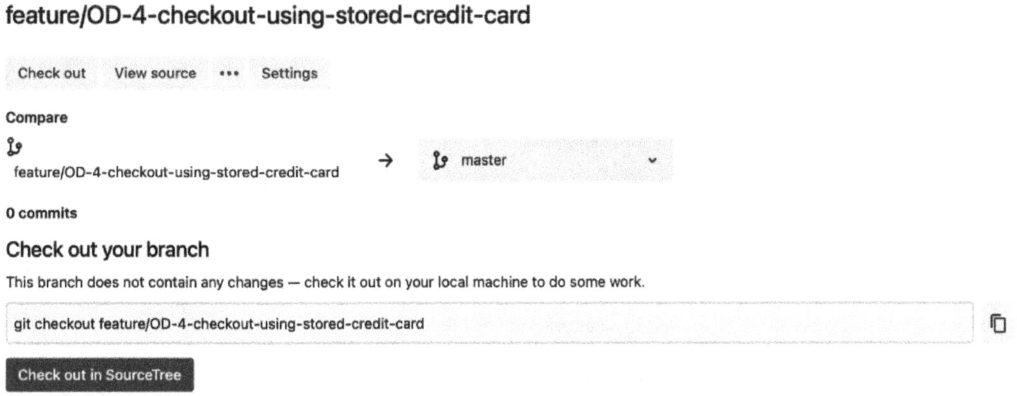

feature/OD-4-checkout-using-stored-credit-card

Check out   View source   •••   Settings

Compare

ᛘ
feature/OD-4-checkout-using-stored-credit-card   →   ᛘ master   ⌄

0 commits

Check out your branch

This branch does not contain any changes — check it out on your local machine to do some work.

git checkout feature/OD-4-checkout-using-stored-credit-card   ⎘

Check out in SourceTree

Figure 13.13 – Branch page on Bitbucket

With that, you've used the Jira-Bitbucket integration to connect a Jira issue to a Bitbucket branch so that you can start development.

Committing changes in Bitbucket/Start Bitbucket Pipeline Build

At the moment, we're performing implementation in our development process. We make changes in our code and commit those changes in Bitbucket. Once we make a commit against a Bitbucket repository, we want Bitbucket Pipelines to start a CI build that includes any testing and security scans.

Getting ready

Place the code we have in the `Chapter13` folder of this book's GitHub repository into the Bitbucket repository that you created for the previous recipe. This also includes copying the `bitbucket-pipelines.yml` file and making sure it is at the root level of your repository.

How to do it...

We will be making changes directly on the Bitbucket UI, not remotely. Once we make a commit, the Bitbucket pipeline should execute. To put this process in motion, perform the following steps:

1. On the Bitbucket repository page, select **Source** from the repository sidebar.

Figure 13.14 – Selecting source

2. In the **Source** window, open the branch pulldown and select the branch you created in the previous recipe.

master
Files
Filter files
Filter branches
Branches
Tags
master
bugfix/DEVOPS1-4-link-goes-to-the-wrong-page
feature/OD-4-checkout-using-stored-credit-card

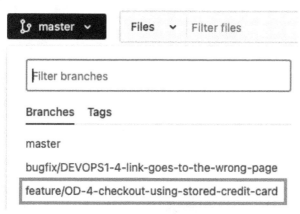

Figure 13.15 – Selecting a branch

3. Select the file you wish to edit. If the file is nested, select the folder first. In this case, we will edit `index.js`.

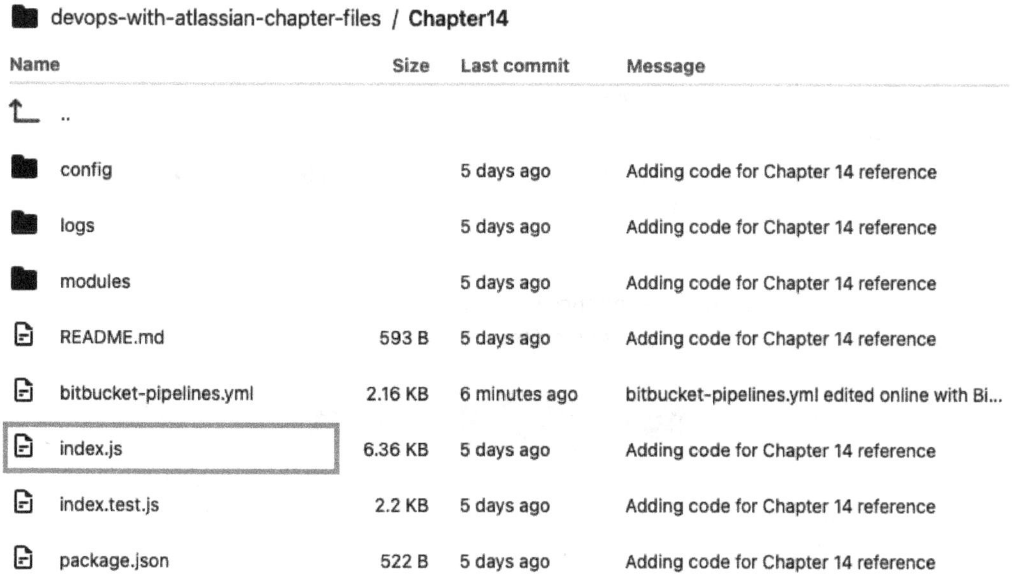

devops-with-atlassian-chapter-files / Chapter14
Name
Size
Last commit
Message
..
config 5 days ago Adding code for Chapter 14 reference
logs 5 days ago Adding code for Chapter 14 reference
modules 5 days ago Adding code for Chapter 14 reference
README.md 593 B 5 days ago Adding code for Chapter 14 reference
bitbucket-pipelines.yml 2.16 KB 6 minutes ago bitbucket-pipelines.yml edited online with Bi...
index.js 6.36 KB 5 days ago Adding code for Chapter 14 reference
index.test.js 2.2 KB 5 days ago Adding code for Chapter 14 reference
package.json 522 B 5 days ago Adding code for Chapter 14 reference

Figure 13.16 – Selecting a file to edit

4.   In the editor page for the `index.js` file, select **Edit**.

Figure 13.17 – Selecting Edit

5.   Make your changes. When your changes are complete, select **Commit** at the bottom of the **Edit** page to add and track your changes in Git.

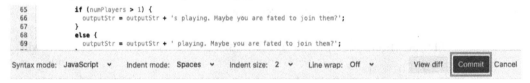

Figure 13.18 – Commit

6.   In the modal, provide a commit message and select **Commit**. For the commit to be linked to Jira, the commit message must start with the issue ID of the relevant Jira issue.

7.   Once the commit occurs, Bitbucket Pipelines should execute, running through the test, scanning, and deployment steps found in `bitbucket-pipelines.yml`. You can view a deployment run and its progress by selecting **Pipelines** in the repository sidebar.

Now that our pipeline is working, let's enhance it by integrating Snyk, which will perform security scanning and move our process more to **DevSecOps**.

Execute Snyk scanning through Bitbucket Pipelines

Snyk is one of several security tools that can be tightly integrated with Bitbucket and Bitbucket Pipelines. This tight integration allows for a Snyk scan to be executed to look for security vulnerabilities as part of testing when it is a step in a Bitbucket pipeline.

Let's learn how to set up this integration.

Getting ready

Integration between Snyk and Bitbucket takes the following forms:

- Snyk is set up as a security provider in Bitbucket so that automated scanning can occur on the repository and its results appear in the **Security** section of the repository sidebar.

- Snyk pipes can execute Snyk scans as part of a Bitbucket Pipelines execution.

Let's look at setting this up.

Setting up Snyk as a security provider

We'll start by setting up Snyk as a security provider for Bitbucket. To do that, perform the following steps:

1. In your Bitbucket repository, select **Security** from the repository sidebar.

Figure 13.19 – Security

2. If you don't have a Snyk account, select **Install Snyk** and set up a Snyk account.

3. A modal will appear, asking you to request access to the entire workspace. Select **Allow** to proceed with integrating Snyk with Bitbucket.

Review permissions

Snyk Security for Bitbucket Cloud requests access to:

- Read your account information
- Read and modify your repositories
- Read and modify your repositories and their pull requests
- Read and modify your repositories' webhooks

Figure 13.20 – Integrating Snyk with Bitbucket

4. Snyk will then ask you to log in or create a new Snyk account. Signing up sets you up for a free trial that moves you to a free plan when the trial period is over. Press **Sign up or Log in** to create or enter your Snyk account.

5. Once you have set up your account and integrated it with Bitbucket Cloud, your repositories will have the following section (annotated) in the repository sidebar:

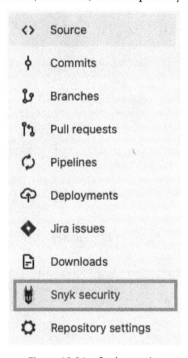

Figure 13.21 – Snyk security

Importing repositories into Snyk and setting up Snyk projects will allow you to perform regular scans and receive notifications regarding any vulnerabilities that have been discovered. We also discussed running Snyk as part of the pipeline by using the Snyk Pipe. Let's learn that's done.

Setting up the Snyk Pipe

We can add a step to perform security scans with Snyk in our `bitbucket-pipelines.yml` file. We originally covered this in *Chapter 6*; we are reinforcing the necessary steps here:

1. Open the `bitbucket-pipelines.yml` file so that you can edit it in the Bitbucket UI. Editing the file should reveal the Bitbucket Pipelines editor, as shown in the following screenshot:

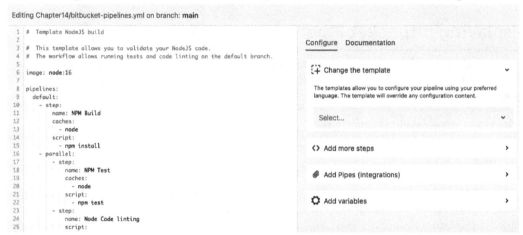

Figure 13.22 – Pipelines editor

2. Add the following code snippet as its own step in the `bitbucket-pipelines.yml` file. To get the Snyk token, go to the **API Token** section of **Account settings** for your Snyk account and save it as a secure repository variable:

```
-pipe: snyk/snyk-scan:1.0.1
    variables:
        SNYK_TOKEN: $SNYK_TOKEN
        LANGUAGE: "node"
```

3. Commit your changes to the `bitbucket-pipelines.yml` file.

With that, Snyk scans have been set up as a step in Bitbucket Pipelines.

Now that Snyk has been configured, it's easy to execute Snyk scans in our pipeline. Let's look into the various methods for doing so.

How to do it...

Now that we've configured Snyk with Bitbucket, let's learn how to execute a Snyk scan:

1. Because any commit will automatically execute the pipeline, place your step in a commonly used branch or the default setting so that a Snyk scan will execute.

2. If you know in which pipeline you placed the Snyk Pipe, you can execute that pipeline manually by selecting the branch, commit, or tag, selecting that pipeline, and clicking **Run**. This was explained in the *Manual execution* recipe in *Chapter 6*.

Next, we'll consider various scenarios where if the tests pass, we move from CI to CD.

Displaying Bitbucket Pipeline Build Status in Compass

Compass organizes development into components and displays the status of these components. Events that affect these components include builds and deployments. Compass can gather information from Jira and Bitbucket and present it in a single interface.

In this recipe, we will make changes in Bitbucket and see their effects in Compass.

Getting ready

If you haven't already done so, you need to connect Bitbucket to Compass. Instructions for doing so can be found in the *Integrating Compass with Bitbucket Cloud* recipe in *Chapter 11*. Another way administrators can connect Bitbucket to Compass comes about when initially installing Compass. Let's see what can be done at this time:

1. Compass identifies if Bitbucket is part of the cloud organization. If it finds Bitbucket repositories, it will offer to map them as Compass components. Verify that Compass has found the desired Bitbucket project.

Figure 13.23 – Importing repositories as Compass components

2. At this point, you will have to allow Compass to connect with Bitbucket. To do so, click the **Grant access** button.

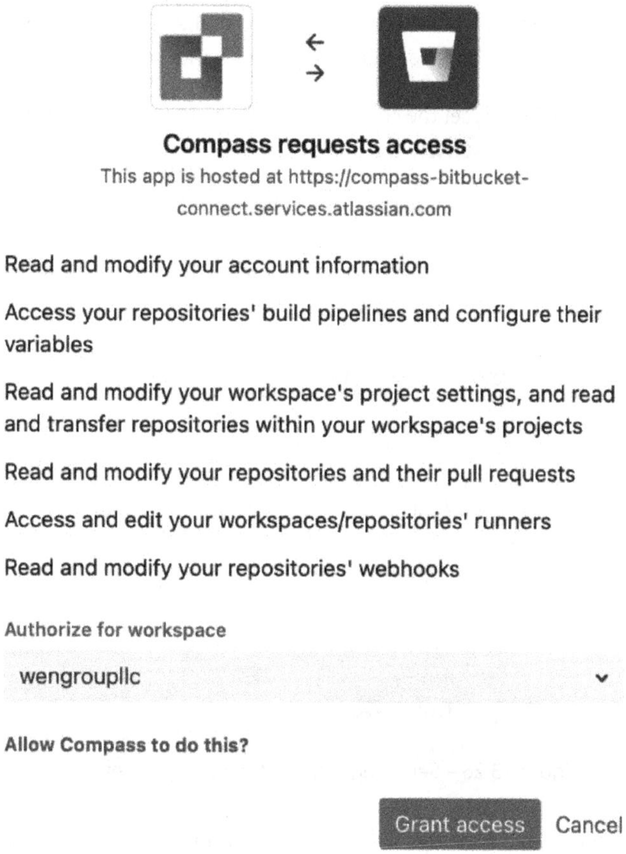

Compass requests access

This app is hosted at https://compass-bitbucket-connect.services.atlassian.com

Read and modify your account information

Access your repositories' build pipelines and configure their variables

Read and modify your workspace's project settings, and read and transfer repositories within your workspace's projects

Read and modify your repositories and their pull requests

Access and edit your workspaces/repositories' runners

Read and modify your repositories' webhooks

Authorize for workspace

wengroupllc   ⌄

Allow Compass to do this?

Grant access   Cancel

Figure 13.24 – Authorizing Compass

Once authorization has been completed, you can view information about your Bitbucket repositories since they've been mapped as Compass components.

2 components

Component	Description	Tier	Repository
DevOps1	Imported from a Bitbucket repository linked to the ...	TIER 4	DevOps1
Component1	Imported from a Bitbucket repository linked to the ...	TIER 4	Component1

Figure 13.25 – Compass components

How to do it...

Now that our Bitbucket repositories have been mapped as Compass components, let's see what happens in Compass when we make changes in Bitbucket. For this visualization, it may be beneficial to have Compass and Bitbucket on separate browser tabs or browser windows:

1. In the **Compass** tab or window, set the component that corresponds to your Bitbucket repository. Click **Activity** to view recent activity for the component.

Figure 13.26 – Selecting Activity for a component

2. In the **Bitbucket** tab, make a change in the repository. Finalize this by clicking the **Commit** button in the editor.

3. In the modal, enter a commit message and other required information. Click **Commit** to commit the change.

    In Bitbucket, go to the **Pipelines** view to confirm that a pipeline has been executed.

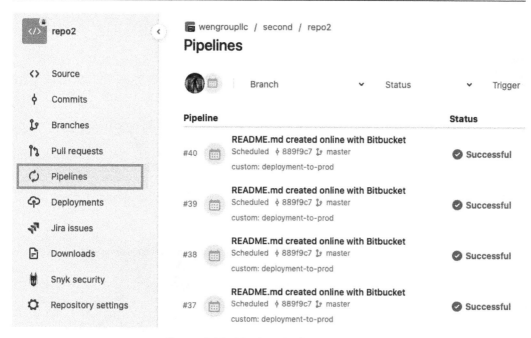

Figure 13.27 – Viewing pipeline executions

4. Go to the browser tab where Compass is running. On the **Activity** view of the component that's been mapped to the Bitbucket repository, confirm that the deployment occurred.

Time	Status	Event	Description
Today, 7:42 PM	✅ Successful	Deployment #5	sample.txt edited online with Bitbucket
Today, 7:22 PM	✅ Successful	Deployment #4	Merged bugfix/DEVOPS1-4-link-goes-to-the-wrong-page int...
Today, 3:02 PM	✅ Successful	Deployment #3	DEVOPS1-4 edited online with Bitbucket - corrected code

Figure 13.28 – Deployments seen in Compass

With that, we've confirmed that actions in Bitbucket will show up in Compass. This information includes deployments.

In our sample scenario, our development team will get the same notifications as operations on any alerts. Let's see how to do that.

Creating an Opsgenie alert for the Jira project

In the *Setting up Opsgenie teams* recipe in *Chapter 12*, we looked at setting up Opsgenie. This allows us to create teams and notification schedules that will react when an alert occurs.

Getting ready

If you didn't connect your Jira project to Opsgenie in *Chapter 12* or have created a new Jira project, follow these steps:

1. In your project, select **Project settings** and then select **Toolchain**.

2. On the **Toolchain** page, expand the **Operate** section and select + **Add on-call team**.

Figure 13.29 – Add on-call team

3. In the modal, type in the name of your team. As you type, the action will appear as a dropdown with the team name you enter. Select the **Create...** drop-down option that appears below. After the new team name is visible as a dropdown, as shown in the following screenshot, click **Create**.

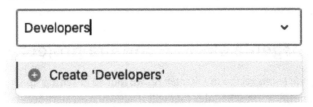

Figure 13.30 – Add on-call team

4. Hover over the team you created. Click the more actions (**...**) icon and select **Open in Opsgenie**.

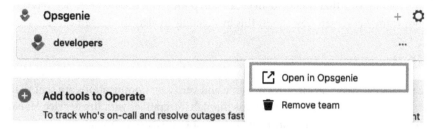

Figure 13.31 – Open in Opsgenie

5. Opsgenie will open on the team page. Look at the **On-call** section and verify the following:

I. The routing rule points to an escalation policy. If it doesn't, refer to the *Escalation policies and rules* recipe in *Chapter 12* to create an escalation policy and set a routing rule. An example is shown in the following screenshot:

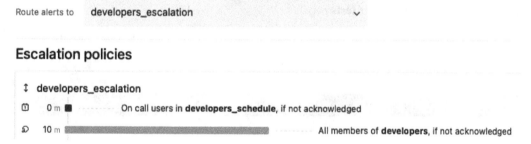

Figure 13.32 – Routing rules and Escalation policies

II. A team member has been identified in **On-call schedules**. This is shown in the following screenshot:

Figure 13.33 – On-call schedules

If you have members on your team, and your team has an on-call schedule with team members identified and a routing rule that points to an escalation policy, you should be all set to receive alerts directed to your team. Let's learn how to test this by creating an alert in Opsgenie.

How to do it...

We are going to directly create an alert in Opsgenie and verify that notifications are sent to the team member on-call. Normally, Opsgenie receives alerts through integrations with **Jira Service Management** or DevOps monitoring tools. We are performing this test to verify our simulation.

To create the alert, perform the following steps in Opsgenie:

1. In Opsgenie, go to the **Alerts** item in the menu bar.

Figure 13.34 – Alerts

2. Select **Create alert** on the **Alerts** page.

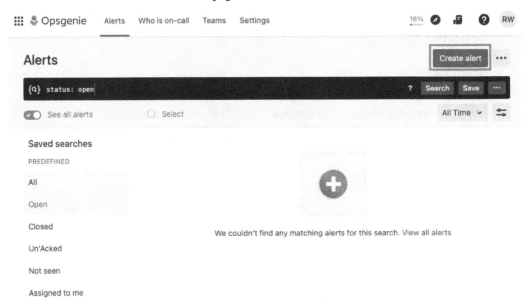

Figure 13.35 – Create alert

3. In the modal, select **Default API** for **API integration** and search for your team in the **Responders** area. Once you have selected these options, fill in **Alert message** and click **Create**.

Figure 13.36 – Setting up and creating a new alert

4. Verify that the alert has been created in Opsgenie.

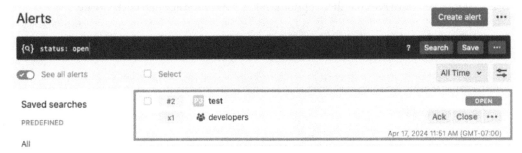

Figure 13.37 – Alert in Opsgenie

5. You can also go to the email tied to the team member who's on-call and verify that an email notification has been sent.

Figure 13.38 – Alert email notification

As we can see, developers can receive notifications about incidents as well as from operations people, allowing developers and operations to work together on production problems. If the root cause is found to be a bug, developers can add fixes by creating bugfix branches for development. We'll see how that process may occur in the next recipe.

Creating a bugfix branch

A problem has come up! QA has discovered a bug in the new release. They wrote a bug in Jira that has been assigned to you. What you have to do is create a branch in Bitbucket that will contain the solution and connect it to the Jira issue.

We will simulate the preceding scenario in this recipe. The first thing you must do is create the bug in Jira.

Getting ready

Our first stop is to use Jira to create the bug. To do this, perform the following steps:

1. Select the **Create** button at the top of the Jira screen.

2. In the modal that appears, fill in mandatory fields such as **Summary** and make sure the **Project** and **Issue type** (**Bug**) fields are correct. Click **Create**.

   Your new bug should appear in the project.

Figure 13.39 – New Bug

With the bug created in Jira, it's time to create the bugfix branch in Bitbucket. Let's examine that further.

How to do it...

Continuing with the same Jira project tied to the Bitbucket repository provided in the *Creating a code change in Bitbucket* recipe, we can easily create our bugfix Bitbucket branch by performing the following steps:

1. In our Jira bug, go to the **Development** section and select **Create branch**.

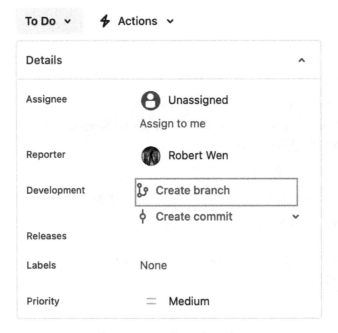

Figure 13.40 – Create branch

2. A page in Bitbucket will appear as a new browser tab or window. It will have the following information filled in:

    - The branch name prefilled with the ID of the Jira issue

    - The parent branch

    - The type of branch

    - The repository name

    Ensure this information is correct and click **Create**.

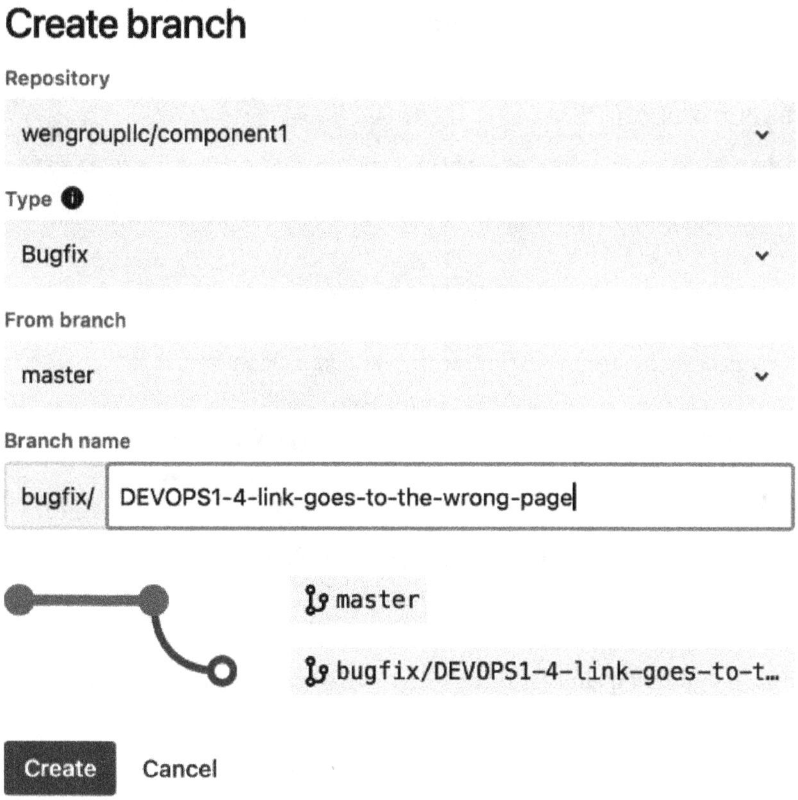

    ## Create branch

    Repository

    wengroupllc/component1   ∨

    Type ⓘ

    Bugfix   ∨

    From branch

    master   ∨

    Branch name

    | bugfix/ | DEVOPS1-4-link-goes-to-the-wrong-page |

    ⎇ master

    ⎇ bugfix/DEVOPS1-4-link-goes-to-t...

    Create   Cancel

    Figure 13.41 – Create branch

3.   The branch should appear in the **Branches** view of your repository, with available actions to check out onto a local environment and an area to **View source**.

bugfix/DEVOPS1-4-link-goes-to-the-wrong-page

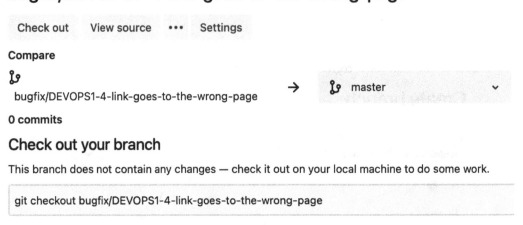

Figure 13.42 – Viewing the branch

With a branch created, we can begin development. Let's take a closer look at this process.

Committing a bugfix and watching the pipeline's execution

We have been diligently troubleshooting for the root cause of our bug and we have developed a fix. What happens when we commit that fix to the repository? We want testing and scanning to occur to ensure high-quality code.

This requires that Bitbucket Pipelines is ready to execute when the commit occurs. Let's learn how to make that possible.

Getting ready

The prerequisites for this recipe are identical to those for the *Committing changes in Bitbucket/Start Bitbucket Pipeline Build* recipe – that is, we need to make sure there is a `bitbucket-pipelines.yml` file in our repository and that it is at the root level of our repository's directory structure.

How to do it...

Just as with the *Committing changes in Bitbucket/Start Bitbucket Pipeline Build* recipe, we are going to observe the pipeline execute by performing the following steps:

1. In our bugfix branch, we must make a change in our file. We can commit our change by clicking **Commit** in our editor.

2. In the modal, enter a value in the **Commit message** field. To have the commit recorded in our Jira issue, the message needs to start with the ID of our Jira issue. Make the necessary changes and click **Commit**.

3. Go to the **Pipelines** view. The pipeline should be executing based on the most recent commit.

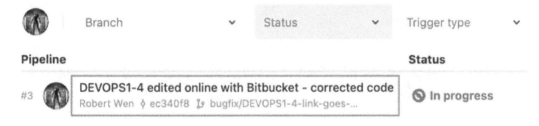

Figure 13.43 – The Pipelines view

4.   Click on the line corresponding to the pipeline's execution to see the pipeline execution details.

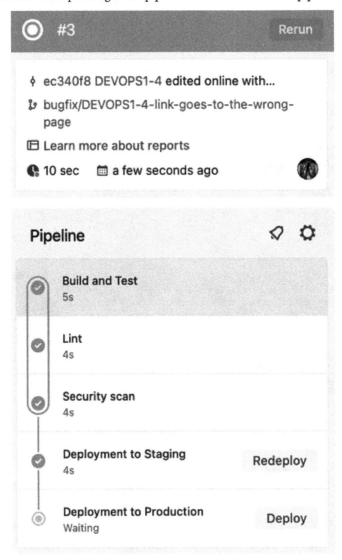

Figure 13.44 – Pipeline execution details

With that, we've learned how to execute a pipeline based on a commit from a bugfix branch.

14

Appendix – Key Takeaways and the Future of Atlassian DevOps Tools

Throughout this book, we have demonstrated how to easily apply integrations to core Atlassian tools and create a DevOps toolchain. But this often raises the question, "Is this it?"

The good news is that Atlassian is continually adding new features to its cloud products that we have showcased throughout this book in previous chapters. Many of these features allow for easy interconnection between Atlassian products to improve visibility and status tracking.

In this chapter, we will focus on the immediate future of Atlassian products and your use of these products. We will center our focus on the following topics:

- Atlassian Intelligence
- Best practices for setting up your toolchain

Let's begin our examination of the future with a look at Atlassian's application of **artificial intelligence (AI)** and **machine learning (ML)**.

Atlassian Intelligence and other AI technologies

Atlassian first introduced **Atlassian Intelligence** in April 2023, and it has changed the way users use Atlassian tools. Not all tools have Atlassian Intelligence enabled, but the main products – Jira, Confluence, and Jira Service Management – have exciting new features that make it easier to use these tools.

Starting off with Jira, Atlassian Intelligence makes it easier to find issues. Prior to Atlassian Intelligence being available in Jira, users could create complex **Jira Query Language (JQL)** searches to find issues of interest. Now, with Atlassian Intelligence, Atlassian allows you to use **natural language (NL)** to simply ask Jira about issues you are interested in. Then, Atlassian Intelligence will write the JQL search on your behalf and return the relevant results.

Atlassian Intelligence does not stop at helping you with your search queries. As you are creating issues, you can leverage the power of Atlassian Intelligence to help you create your issues. Atlassian Intelligence can help you brainstorm ideas and help populate a description of your issues. This can be very helpful whenever you need a little nudge or sprinkle of inspiration to get going.

Finally, Atlassian Intelligence helps you change the tone of your writing. The tone is sometimes difficult to convey in written form, and the primary way users interact with Jira is through written words. This can lead to conflicts, confusion, and disagreements. Luckily, Atlassian Intelligence helps users improve their writing by rewriting text into different types of tones.

Confluence also has some very exciting new features that are powered by Atlassian Intelligence. Similar to Jira's JQL, Confluence also has a powerful search capability that allows users to find data. Historically, data searches through Confluence pages and relevant pages were returned back to the user. Now, with the power of Atlassian Intelligence, not only are relevant pages returned, but an answer to the user's query is returned as well. This changes the way users search for information within Confluence because instead of looking for keywords and hoping that they find some information, Atlassian Intelligence builds an answer for you. The best part is that a user can use NL to ask their question in Confluence instead of relying on keywords.

Another Atlassian Intelligence feature in Confluence is the ability to help you look up words. Many teams have technical terms that are specific to their team or company. These technical terms are typically common knowledge for team members, but for new team members, it can be challenging to keep up with the definitions of all these terms. Luckily for these new employees, Atlassian Intelligence can easily search through all available Confluence data to help them figure out what a term means. This is great because it doesn't go out to the internet to find what the term means but rather sources that data from the company's personal data, which increases the likelihood that a relevant definition is found.

One last Atlassian Intelligence feature worth noting in Confluence is the ability to summarize Confluence pages for a user. Pages tend to contain a lot of data, and it can be hard to fully digest all the information contained within a page. With Atlassian Intelligence, a user can receive a summarized view of the page, highlighting key pieces of information that a user should know about.

Atlassian Intelligence is still fairly new, and Atlassian continues to improve it. Keep an eye out as Atlassian continues to expand the capabilities of Atlassian Intelligence and spread these to other products in the Atlassian suite.

Best practices for setting up your toolchain

As you have learned in this book, setting up all the tools and getting them all configured is an involved process. At this point, you should have all the tools configured, and your DevOps tools should be ready for your team to increase their collaboration and efficiency. Next are some best practices that you should consider when setting up your toolchain to ensure a smoother time setting things up:

- **Define clear objectives**: Before you start deploying your DevOps tools, figure out what it is you want to achieve. Many different DevOps tools do almost the exact same thing, and picking between them might not be trivial. This book focuses on Atlassian tools, but there are many different alternatives. If your company's objective was to stay within the Atlassian ecosystem, then this book set you up to meet that objective.

- **Select the right tools**: With a clear objective in mind, you need to select the right set of tools. As mentioned before, there are many DevOps tools that are very similar in price and functionality. Evaluate each tool, understanding the pros and cons of each tool with respect to what you and your team need. Avoid signing up multiple tools that end up doing the same thing because this only adds complexity to your DevOps tools and increases the overall spend and inefficiency of your team.

- **Standardize your workflow**: It can be tempting to create a unique workflow or process for each of your developer teams. Ideally, your company should have a well-defined process on how to utilize your DevOps tools. Avoid having unnecessary steps or steps that could create confusion for your team. Working out of the same operating model can help increase team collaboration, communication, and effectiveness since they will all speak a common language. Make sure your process is well documented and as you onboard new employees, they become familiar with your process. Refine your process regularly as you make new discoveries or get more efficiency out of your tools and processes.

- **Automate the process**: Automate as much as possible; this doesn't mean automate everything. Automate parts of your process and toolchain that make sense. Don't automate for the sake of automating. Having a human in the loop may be required until your team or process matures over time. Don't try to automate everything on day one if your team doesn't fully understand how to do the work manually first.

- **Track the performance**: Monitor and measure the performance of your DevOps tools and processes. There is always room for improvement, and regularly revisiting your tools and processes can help increase your team's effectiveness and throughput. Establish some metrics around **key performance indicators** (**KPIs**) that will guide your team through their DevOps journey.

- **Maintain security and compliance**: Regularly review your configurations and processes to ensure that your team is keeping up with the latest trends and threats in the cyber security space. Review your code review practices, vulnerability scans, and access control policies to make sure that your team is always adapting to the ever-changing cyber landscape.

- **Collaborate and communicate**: Having great DevOps tools is fantastic, but no matter how great your toolchain is, make sure you encourage your team to collaborate and communicate. Your DevOps tools are there to help your team, but they should not be there to replace your team's interactions with each other.

- **Training and development**: Finally, train your team on how to use your tools. DevOps can be confusing for your team, especially new team members who might not be too familiar with your infrastructure. Get in the habit of training new members and providing refresher training for existing employees. As you make changes and improvements to your process and toolchain, remember to update your documentation and, if needed, share those changes with the rest of your team. Encourage a culture of sharing new discoveries or better ways of using your tools.

Summary

In conclusion, DevOps is more of a journey and less of a destination. This book was designed to get you started and help you jumpstart your DevOps toolchain. As your team continues to utilize the tools that were configured in this book, you should continue to improve upon your toolchain and defined processes. In this book, we explored the entire DevOps lifecycle and covered configurations for Jira, Jira Product Discovery, Bitbucket, Compass, Opsgenie, and much more. This. is only the beginning. Your team will naturally continue to tweak and make changes to your configurations.

This book has covered how important and valuable it is to choose the right set of DevOps tools. The Atlassian tool suite is designed to work well together, and any team using Jira or Confluence should highly consider folding in the rest of the Atlassian tools.

As we conclude this book, we want to remind you that this is only the beginning. Now that you have a functioning set of tools, it is time for your team to really leverage the power of DevOps to take your software quality and software development efficiencies to the next level. DevOps is about continuously learning and improving, and your tools should follow the same practice.

We hope that this book has provided you with valuable insights, knowledge, and inspiration to lead your team's DevOps efforts. Thank you for joining us on this journey, and we wish you all the best!

Index

packtpub.com

Subscribe to our online digital library for full access to over 7,000 books and videos, as well as industry leading tools to help you plan your personal development and advance your career. For more information, please visit our website.

Why subscribe?

- Spend less time learning and more time coding with practical eBooks and Videos from over 4,000 industry professionals

- Improve your learning with Skill Plans built especially for you

- Get a free eBook or video every month

- Fully searchable for easy access to vital information

- Copy and paste, print, and bookmark content

Did you know that Packt offers eBook versions of every book published, with PDF and ePub files available? You can upgrade to the eBook version at packtpub.com and as a print book customer, you are entitled to a discount on the eBook copy. Get in touch with us at customercare@packtpub.com for more details.

At www.packtpub.com, you can also read a collection of free technical articles, sign up for a range of free newsletters, and receive exclusive discounts and offers on Packt books and eBooks.

Other Books You May Enjoy

If you enjoyed this book, you may be interested in these other books by Packt:

Modern DevOps Practices

Gaurav Agarwal

ISBN: 978-1-80512-182-4

- Explore modern DevOps practices with Git and GitOps
- Master container fundamentals with Docker and Kubernetes
- Become well versed in AWS ECS, Google Cloud Run, and Knative
- Discover how to efficiently build and manage secure Docker images
- Understand continuous integration with Jenkins on Kubernetes and GitHub Actions
- Get to grips with using Argo CD for continuous deployment and delivery
- Manage immutable infrastructure on the cloud with Packer, Terraform, and Ansible
- Operate container applications in production using Istio and learn about AI in DevOps

The Ultimate Docker Container Book

Dr. Gabriel N. Schenker

ISBN: 978-1-80461-398-6

- Understand the benefits of using containers
- Manage Docker containers effectively
- Create and manage Docker images
- Explore data volumes and environment variables
- Master distributed application architecture
- Deep dive into Docker networking
- Use Docker Compose for multi-service apps
- Deploy apps on major cloud platforms

Packt is searching for authors like you

If you're interested in becoming an author for Packt, please visit `authors.packtpub.com` and apply today. We have worked with thousands of developers and tech professionals, just like you, to help them share their insight with the global tech community. You can make a general application, apply for a specific hot topic that we are recruiting an author for, or submit your own idea.

Share Your Thoughts

Now you've finished *Atlassian DevOps Toolchain Cookbook*, we'd love to hear your thoughts! Scan the QR code below to go straight to the Amazon review page for this book and share your feedback or leave a review on the site that you purchased it from.

`https://packt.link/r/1835463789`

Your review is important to us and the tech community and will help us make sure we're delivering excellent quality content.

Download a free PDF copy of this book

Thanks for purchasing this book!

Do you like to read on the go but are unable to carry your print books everywhere?

Is your eBook purchase not compatible with the device of your choice?

Don't worry, now with every Packt book you get a DRM-free PDF version of that book at no cost.

Read anywhere, any place, on any device. Search, copy, and paste code from your favorite technical books directly into your application.

The perks don't stop there, you can get exclusive access to discounts, newsletters, and great free content in your inbox daily

Follow these simple steps to get the benefits:

1. Scan the QR code or visit the link below

https://packt.link/free-ebook/978-1-83546-378-9

2. Submit your proof of purchase

3. That's it! We'll send your free PDF and other benefits to your email directly